Mitochondrial Physiology and Pathology

Van Nostrand Reinhold Advanced Cell Biology Series
Miguel M. Azar, Series Editor

CONTROL OF CELL GROWTH AND PROLIFERATION, edited by Carlo M. Veneziale
NEW METHODOLOGIES IN STUDIES OF PROTEIN CONFIGURATION, edited by Tai Te Wu
CHROMOSOME STRUCTURE AND FUNCTION, edited by Michael S. Risley
MITOCHONDRIAL PHYSIOLOGY AND PATHOLOGY, edited by Gary Fiskum

Additional volumes in preparation

Mitochondrial Physiology and Pathology

GARY FISKUM, Editor
George Washington University
School of Medicine

Van Nostrand Reinhold
Advanced Cell Biology Series

VNR VAN NOSTRAND REINHOLD COMPANY
New York

Library of Congress Catalog Card Number: 86-5485
ISBN: 0-442-22725-6

Manufactured in the United States of America.

Published by Van Nostrand Reinhold Company Inc.
115 Fifth Avenue
New York, New York 10003

Van Nostrand Reinhold Company Limited
Molly Millars Lane
Wokingham, Berkshire RG11 2PY, England

Van Nostrand Reinhold
480 Latrobe Street
Melbourne, Victoria 3000, Australia

Macmillan of Canada
Division of Gage Publishing Limited
164 Commander Boulevard
Agincourt, Ontario MIS 3C7, Canada

15 14 13 12 11 10 9 8 7 6 5 4 3 2 1

Library of Congress Cataloging-in-Publication Data
Mitochondrial physiology and pathology.
(Van Nostrand Reinhold advanced cell biology series)
Includes index.
1. Mitochondria. 2. Pathology, Cellular.
I. Fiskum, Gary. II. Series. [DNLM: 1. Mitochondria—physiology. 2. Mitochondria—ultrastructure.
QH 603.M5 M68425]
QH603.M5M565 1986 574.87′342 86-5485
ISBN 0-442-22725-6

In memory of
Albert L. Lehninger
(1917–1986)

Contents

Foreword

We take great pleasure in presenting *Mitochondrial Physiology and Pathology* to the scientific community. We know the research to be timely, pertinent, and of very high quality. The editor, Dr. Gary Fiskum, is a distinguished faculty member in the department of biochemistry at George Washington University School of Medicine, Washington, D.C.

To many, mitochondria are simply the "power plants" of the cell. Dr. Fiskum and collaborators tell us of the many cellular functions in which mitochondria participate and how these activities affect normal and abnormal cell physiology. Aspects of metabolic injury, tumoral cells, fetal life, and its proper biological relevance are also discussed. We hope the readers will share the contributors' views on these highly dynamic and heterogeneous organelles.

It is our aim that scientists following our Advance Cell Biology Series share and extend our optimism and learning experience as we participate in the preparation of this monograph.

MIGUEL M. AZAR, M.D., Ph.D.

Preface

Research in the biology of mitochondria has expanded far beyond the confines of bioenergetics and metabolism to include multidisciplinary fields such as molecular genetics, endocrinology, and pathophysiology. The different perspectives offered by investigators in these fields together with the development of improved techniques for measuring mitochondrial activities in situ have broadened our understanding of the roles mitochondria play in normal and abnormal cell physiology.

In *Mitochondrial Physiology and Pathology*, leading researchers review recent advances in physiological aspects of mitochondrial function. Four chapters are devoted to normal mitochondrial activity including protein turnover, calcium transport, heat production in brown fat tissue, and the metabolic changes that accompany perinatal development. The remaining chapters cover several aspects of abnormal mitochondrial function including those associated with cerebral and cardiac ischemia, malignant transformation, and the cytotoxic effects of activated macrophages toward neoplastic cells. This unique collection of reviews will be of particular interest to investigators involved in the areas of membrane biology, cell physiology, calcium homeostasis, energy metabolism, and biochemical pathophysiology.

GARY FISKUM

Contributors

June R. Aprille
Department of Biology, Tufts University, Medford, Massachusetts 02155

Michel Desautels
College of Medicine, University of Saskatchewan, Saskatoon, Saskatchewan, Canada S7N OWO

Gary Fiskum
Department of Biochemistry, George Washington University School of Medicine, Washington, D.C. 20037

Donald L. Granger
Department of Medicine, Duke University Medical Center, Durham, North Carolina 27706

John B. Hibbs, Jr.
Department of Medicine, University of Utah Medical Center, Salt Lake City, Utah 84112

Lars Hillered
Department of Neurology, University of California School of Medicine, San Francisco, California 94143

Kathryn F. LaNoue
Department of Physiology, The Milton S. Hershey Medical Center, The Pennsylvania State University, Hershey, Pennsylvania 17033

Albert L. Lehninger (deceased)
Department of Biochemistry, Johns Hopkins University Medical Center, Baltimore, Maryland 21205

Jeanie B. McMillin-Wood
Department of Medicine, Baylor College of Medicine, Houston, Texas 77030

Elizabeth Murphy
National Institute of Environmental Health Sciences, Research Triangle Park, North Carolina 27709

Paul E. Wolkowicz
Department of Medicine, Baylor College of Medicine, Houston, Texas 77030

Mitochondrial Physiology and Pathology

1

Regulation of Energy Coupling in Brown Fat

Kathryn F. LaNoue

The Pennsylvania State University

INTRODUCTION AND OVERVIEW

Brown fat is uniquely interesting tissue for studies of energy metabolism because respiration rates of brown fat are not coupled to energy demand as they are in other mammalian tissues (Nicholls and Locke, 1984; Himms-Hagen, 1976). Normally, rates of oxidation, and thus ATP production, are geared to the cell's requirements for ATP-utilizing processes such as muscle contraction, ion pumping, biosynthesis, and cell growth (Crass, McCaskill, and Shipp, 1969; Oram, Bennetch, and Neely, 1973; Balaban et al., 1980). The precise molecular mechanisms that mediate the phenomenon of respiratory control are not known but in most cells respiration appears to be a function of cytosolic ATP-utilization rates rather than substrate or O_2 availability.

The major function of brown adipose tissue is to generate heat acutely in periods of cold stress (Nicholls and Locke, 1984). In certain species (Rothwell and Stock, 1979, 1980; Rothwell, Stock, and Stribling, 1982) brown fat may also serve the function of burning excess calories to prevent obesity. The sympathetic nervous system provides rapid communication between the brain and brown fat deposits; the sympathetic nerve terminals release norepinephrine within the brown fat (Himms-Hagen, 1970) and then norepinephrine binds to β-receptors on the surface of the cell membrane (Bukowiecki et al., 1980; Svoboda et al., 1979). Cyclic AMP levels in the cells increase within seconds (Pettersson and Vallin, 1976; Skala and Knight, 1979) and this increase is followed by a rapid lipolysis of triglycerides (Nedergaard and

Lindberg, 1979; Fain, Jacobs, and Clement-Cormier, 1973). Simultaneously or soon thereafter, respiration rates within the tissue increase up to 40-fold (Fain, Jacobs, and Clement-Cormier, 1973) and blood flow increases at least 20-fold (Foster and Frydman, 1978*a*, 1978*b*, 1979). However, no comparable increase in cytosolic ATP hydrolysis has been demonstrated (Nicholls and Locke, 1984; Cannon and Vogel, 1977; Bulychev et al., 1972; Horwitz and Easton, 1975). Although much is known about the unique control of respiration in isolated brown fat mitochondria (Locke and Nicholls, 1981; Nicholls and Bernson, 1977), intracellular signals that increase respiration in situ still have not been identified. Also, the mechanisms by which the dramatic changes in blood flow are controlled have not yet been elucidated.

Brown fat is present in significant quantities in infant mammals, in hibernators, and in small mammals that are exposed to cold environments (Lindberg, 1970; Smith and Horwitz, 1969). It exists in discrete and scattered areas of the body. A butterfly-shaped pad between the scapulae is a prominant anatomical feature of those animals whose overall metabolism is significantly affected by the presence of brown fat. In these animals intravenous infusion of norepinephrine raises rates of whole body O_2 utilization four- to five-fold (Himms-Hagen, 1976).

In mammals the brown fat deposits proliferate and the amount of mitochondria per gram of tissue increases when the animal is chronically exposed to cold (Barnard, Mory, and Nechad, 1980; Desautels and Himms-Hagen, 1979). Likewise, rodents who are fed a palatable, high calorie diet exhibit brown fat proliferation and these overfed rodents do not gain weight, but rather burn more calories (Rothwell and Stock, 1979, 1980; Rothwell, Stock, and Stribling, 1982). Significant deposits of brown fat have also been observed in adult human patients with certain forms of adrenal tumors (Ricquier, Nechad, and Mory, 1982; Lean and James, 1983; Chakrabarty, Chaudhure, and Jeffay, 1983). Patients with massive trauma and large burns and therefore high adrenal output frequently pass through a period of hyperphagia without gaining weight (Caldwell, 1962), which may in part be the result of brown fat proliferation.

The mechanisms by which brown fat proliferates while altering its protein and mitochondrial content have not been fully explored (Barnard, Mory, and Nechad, 1980). Likewise, in a variety of species including humans the role that brown fat may play in the overall energy balance is a matter of considerable debate.

Numerous reviews have been written on the subject in recent years. Some of these reviews have provided an overview of brown fat metabolism (Nicholls and Locke, 1984; Himms-Hagen, 1976; Lindberg, 1970; Smith and Horwitz, 1969) and of the mechanisms by which the tissue proliferates (Barnard, Mory, and Nechad, 1980; Rothwell and Stock, 1982). Others have concentrated on the possible role of brown fat in the maintenance of body weight (Rothwell, Stock, and Stribling, 1982; Rothwell and Stock, 1981; Himms-Hagen, 1979). Because of limitations of space, and the specific interests of the author, the present review will focus only on the acute regulation of respiration.

POSSIBLE MECHANISMS FOR ENERGY COUPLING TO HEAT PRODUCTION

In almost all mammalian tissue, oxygen consumption by isolated mitochondria as well as mitochondria in situ is linked stoichiometrically to rates of ATP synthesis. The molar ratio ATP synthesized to oxygen (O_2) consumed is approximately six to one (Balaban et al., 1980; Lehninger, 1975). The two processes are so tightly coupled that oxygen consumption is very slow if ATP synthesis is blocked by lack of substrate or by a specific inhibitor of ATP synthesis such as oligomycin (Nicholls, 1982; Davis and Davis-van Thienen, 1978). Using isolated mitochondria it can be demonstrated that respiration is a function of the ratios of ATP/ADP in the suspending media or of the media phosphorylation potential (ATP)(ADP)/P_i when phosphate levels are lower than 2 to 3 mM (Davis and Davis-van Thienen, 1978; Kunz et al., 1981, 1983). In the intact cell, most energy-consuming processes are linked to the hydrolysis of ATP. When energy consumption is rapid, cytosolic ATP/ADP ratios decrease and respiration increases, restoring cellular ATP homeostasis. The bulk of the evidence suggests that the strict coupling of O_2 consumption to ATP synthesis occurs via dependence of both processes on the electrochemical concentration gradient of protons ($\Delta\mu H^+$) across the mitochondrial inner membrane (Nicholls, 1982). The electron transport chain components pump protons out of the mitochondrial matrix space (Wikstrom, Krab, and Saraste, 1981) whereas the ATP synthase complex, operating in the direction of ATP synthesis allows the flow of protons back into the matrix (Senior and Wise, 1983). Several recent reviews have summarized the evidence in favour of this chemiosmotic hypothesis of energy coupling (Nicholls, 1982; Mitchell, 1976). The basic concepts appear sound, except for some quantitative discrepancies suggesting that a degree of direct interaction between the ATP synthase and electron transport complexes may occur, mediated probably by localized proton currents (Rottenberg, 1983; Ferguson and Sorgato, 1982).

Certain compounds that are capable of uncoupling oxidation from phosphorylation have been identified. These compounds (e.g., dinitrophenol) are now known to be proton ionophores, that is, they can transport protons across natural or artificial lipid bilayers and therefore can replace and/or compete with the ATP synthase in allowing the electrochemical potential of protons built up by the electron transport chain to be dissipated (Mitchell, 1976).

When stimulated by norepinephrine, the rate of oxygen consumption in μmoles O_2/min/mg of cellular protein is higher in brown fat (0.27) (Nedergaard and Lindberg, 1982) than in cardiac tissue (0.12) (Kobayashi and Neely, 1979). Since there are no readily identifiable processes, such as muscle contraction, that will consume ATP at a comparable rate in brown fat, it was initially assumed that in the presence of β-agonists some endogenous compound could uncouple respiration from phosphorylation in the brown adipocytes. The first suggestion that fatty acids could play such a role in metabolism was made in 1956 by Pressman and Lardy. A specific role for free fatty acids as the physiological uncouplers in brown fat was suggested by Fain, Reed, and Saperstein (1967) and by Prusiner, Cannon, and Lindberg (1968). Although the hypothesis that fatty acids generated by the action of a hormone-sensitive lipase can uncouple brown fat has a certain simplistic appeal, the idea was subsequently dropped (until recently), because it became apparent that fatty acids bind to intracellular proteins and it seemed unlikely that they would uncouple mitochondria at the low "unbound" concentration likely to exist intracellularly (Nicholls, 1979; Lindberg, Nedergaard, and Cannon, 1981).

There would perhaps be no necessity for postulating various mechanisms for dissipating the brown fat $\Delta\mu H^+$ if brown fat could oxidize fatty acids without generating a proton gradient. However, the electron transport chain in brown fat mitochondria appears to be similar to the electron transport in all other mitochondria and pumps protons in the same stoichiometric proportions as liver and heart mitochondrial electron transport chains (Nicholls, 1974*a*). Also, the extramitochondrial pathway of fatty acid oxidation present in peroxisomes is not sufficiently active to account for a significant percentage of hormone-stimulated oxygen consumption (Cannon, Alexson, and Nedergaard, 1982). Therefore, a very active means of dissipating the energy generated by the electron transport chain must be sought.

A variety of energy-dissipating *futile* substrate cycles exist in mammalian and other eukaryotic cell types. These cycles have been reviewed extensively (Newsholme and Crabtree, 1976; Newsholme, 1980). Futile cycles are thought to provide an amplification of metabolic control (Newsholme, Challiss, and Crabtree, 1984) and a pathway for regulated thermogenesis (Newsholme and Crabtree, 1976). A futile cycle is defined as one in which metabolite 1 is converted to metabolite 2 by a pathway which involves an input of energy (e.g., a kinase reaction). Simultaneously, a second enzyme converts metabolite 2 back to metabolite 1, without regeneration of the original energy (e.g., a phosphatase). The net effect of the cycling is hydrolysis of ATP and generation of heat. In its most general form, futile cycling can refer to ion cycling as well. Thus, Na^+ can be pumped out of the cell by Na^+, K^+-ATPase and then returned to the cell cytosol via a regulated Na^+ channel. The net result is simply hydrolysis of ATP and heat generation, while the regulated nature of the channel may provide for a variety of interesting control and signal possibilities. Regulation of Na^+ cycling is thought to be important in the mechanism of action of thyroid hormones for stimulating basal metabolic rates (Edelman, 1974). Na^+ cycling was thought to be involved in stimulating brown fat respiration (Horwitz, 1979) but now it appears that proton cycling through the mitochondrial membrane is more important than Na^+ cycling through the plasma membrane (Nicholls and Locke, 1984).

A regulated channel for protons called thermogenin (Cannon, Nedergaard, and Sundin, 1981) or uncoupling protein (Nicholls and Locke, 1984) has been identified in the inner mitochondrial membrane of brown fat mitochondria. This uncoupling proton cannot be found in other mitochondria. Evidence to be outlined next suggests that proton cycling out of the mitochondrial matrix via the proton pumps of the electron transport chain and back into the mitochondria, via the uncoupling protein, is responsible for the majority of the heat generated in brown fat. Uncertainties about the physiological mechanism of regulation of the uncoupling protein remain an area of current research. Also it appears to this reviewer that a very active pathway of substrate cycling, not yet identified, must exist in brown fat to hydrolyze the considerable amount of ATP that is generated by the succinyl CoA thiokinase reaction in the citric acid cycle. Evidence in favor of the three types of futile cycles likely to be present in brown fat—Na^+ cycling, H^+ cycling, and substrate cycling—will be discussed next as well as the maximal capacities of these cycles and the mechanisms that may control them.

Na^+ CYCLING IN BROWN FAT CELLS

By 1968 the thermogenic function of brown fat was well recognized. Fatty acids were considered as possible uncoupling agents, but substrate cycling via re-esterification of fatty acids had been shown to be too slow to account for the increased O_2 consumption (Lindberg et al., 1967; Prusiner et al., 1968). Girardier, Seydoux, and Clausen (1968) suggested that ATP hydrolysis via the Na^+, K^+-ATPase in the plasma membrane of brown adipocytes could account for the increased rates of respiration. Using microelectrodes to measure cell membrane potentials on surface cells of excised, perifused, brown fat pads in rats, they found that norepinephrine produced a rapid, large decrease in the electrical potential of the outer cell membrane ($\Delta\Psi_c$) from ~50 mvolts (negative inside) to ~25 mvolts (negative inside). The potential decrease was sustained while norepinephrine remained in the perifusate. Studies of the effects of varying internal K^+ on the potential suggested that the ratio of the permeability of the Na^+ to K^+ was considerably higher in brown fat than in electrically excitable tissues such as muscle and nerve. In muscle and nerve the value of $\Delta\Psi_c$ remains very near the K^+ diffusion gradient and is insensitive to the gradient of Na^+. The relatively high permeability of Na^+ to K^+ was confirmed in later studies (Williams and Matthews, 1974*a,* 1974*b*). However, the observation that ouabain, a potent inhibitor of Na^+, K^+-ATPase decreased the potential by only 50% over a 2-hour period suggested a rather low overall ion permeability. Workers (Williams and Matthews, 1974*a;* Fink and Williams, 1976) have subsequently found somewhat shorter depolarization half times (0.5 hours), but our own studies (LaNoue et al., 1986) of Na^+ gradients in the presence of ouabain tend to support the earlier report. Surprisingly, anaerobiosis caused a rapid depolarization (Girardier, Seydoux, and Clausen, 1968) very similar to that observed due to the presence of norepinephrine. Girardier, Seydoux, and Clausen (1968) concluded that norepinephrine and anaerobiosis produced a large increase in the Na^+ permeability, that the Na^+, K^+-ATPase was stimulated by the increased cytosolic Na^+, and that excess ADP generated by the activity of the Na^+ pump caused the increase in respiration.

Subsequent studies over the next few years tended to substantiate the hypothesis. The decrease in $\Delta\Psi_c$ in excised tissue blocks was confirmed. Surprisingly, it was found that elimination of $(HCO_3)^{-1}$ (Williams and Matthews, 1974*a*) from the incubation media depolarized the cells, that iodoacetate would slowly depolarize the cells, but that CN^- did not mimic the effect of anaerobiosis (Williams and Mathews, 1974*a*). However, when experiments were carried out with intact cold-adapted rats in order to measure the $\Delta\Psi_c$ of unexcised adipocytes in situ, the decrease in $\Delta\Psi_c$ was found to be smaller (8-10 volts) and strikingly transient, since the cells repolarized after only a few minutes (Horwitz, Horowitz, and Smith, 1969; Horowitz, Horwitz, and Smith, 1971; Flaim, Horwitz and Horowitz, 1977). The transient depolarization was accompanied by a similarly transient increase in conductivity across the cell membranes (Horowitz, Horwitz and Smith, 1971). In retrospect it seems possible that the transient depolarization was caused by a transient hypoxia, reversed rather quickly by the stimulation of blood flow through the in situ tissue. At that time, however, other types of experiments lent support to the concept of direct norepinephrine stimulation of Na^+ cycling. Thus, ouabain was observed to partially inhibit the increase in respiration caused by norepinephrine (Herd, Hammond, and Hamolsky, 1973; Horwitz, 1973) and several laboratories reported a stimulation by norepinephrine of the Na^+, K^+-ATPase in membrane vesicle preparations (Horwitz and Easton, 1975; Rothwell, Stock, and Wyllie, 1981; Hettinger and Horwitz, 1983).

However, it has become increasingly obvious in recent years that the Na^+ cycling hypothesis, at least as originally formulated, does not explain the available data. The most frequently stated objection to the Na^+ cycling hypothesis is that neither the Na^+, K^+-ATPase (Horwitz and Easton, 1975; Rothwell, Stock, and Wyllie, 1981; Hettinger and Horwitz, 1983) nor the mitochondrial ATP synthase (Bulychev et al., 1972) have sufficiently high activity to support the respiration rates observed in the presence of norepinephrine. Additionally a relatively recent careful study (Nedergaard, 1981) of the effects of a variety of external ions (Na^+, K^+, Mg^{++}, and Ca^{++}) on the cellular hormonal responses showed that neither hormone sensitive lipolysis nor hormone sensitive respiration responded in a sensitive way to moderate alterations of the external ionic composition.

The rate of respiration of isolated adipocytes at 37°C is 700 natoms O_2/min · 10^6 cells, which, assuming a P/O of 3, suggests a rate of cytosolic ATP hydrolysis of 2.1 μmoles/min · 10^6 cells. The rate of K^+-dependent ATPase in the supernatant fraction of brown fat homogenates is 6 μmoles/hr · mg N (Horwitz and Easton, 1975). There is about 1.5 mg of protein in 10^6 cells, of which 16% is N. Even if all of the cell protein were in the supernatant fraction of the cell homogenate, this amount would translate to only 0.016 μmoles/min · 10^6 cells.

Recently the kinetics of the brown fat Na^+, K^+-ATPase have been re-examined by Horwitz and co-workers (Hettinger and Horwitz, 1983), who have now determined the effect of norepinephrine on the K_m and V_{max} of the ATPase. Vanadium-free ATP was used and experiments were performed with an improved and more highly purified membrane preparation. The V_{max} was lower than previously reported (3 μmoles/hr · mg N) and did not change with norepinephrine. The only effect of norepinephrine was to lower the K_m of K^+ from 8.2 mM to 5.3 mM.

It therefore appears that the effect of ouabain on the norepinephrine stimulation of respiration is due to changes in intracellular ion contents. Only very small effects of ouabain are observed unless the tissue is preincubated with the inhibitor for 15 to 30 minutes (Nicholls and Locke, 1984; Herd, Hammond, and Hamolsky, 1973).

The cell membrane depolarizations caused by norepinephrine have remained of interest, and work has continued in an attempt to elucidate cause-effect relationships. Suggestions have been made that Na^+, K^+-ATPase or a Na^+ channel may be the initial target of hormone action. A small initial drop in cytosolic ATP (LaNoue, Koch, and Meditz, 1982) or change in cytosolic ion levels (Horwitz, 1979) might act as a signaling device, thus triggering later events. However, it now seems likely that the initial observations of hormone-induced decreases in $\Delta\Psi_c$ (Girardier, Seydoux, and Clausen, 1968; Williams and Matthews, 1974*a*, 1974*b*; Fink and Williams, 1976) were due to increases in sodium permeability induced by anaerobiosis rather than directly by hormone as originally suggested (Girardier, Seydoux, and Clausen, 1968). This conclusion is supported by measurements of the tissue's NAD(P)H/

$NAD(P)^+$ ratios in the control and hormonal states. Thus, later studies (Seydoux et al., 1982, 1983; Girardier and Schneider-Picard, 1983) of perifused, excised brown fat pads have included measurement of the cellular $NAD(P)H/NAD(P)^+$ redox states, observed via measurements of surface fluorescence. The data clearly indicate a dramatic increase in the ratio of $NAD(P)H/NAD(P)^+$. This increase could be due to anaerobiosis or simply to provision of excess substrate via the hormone-stimulated lipolysis. It is possible to distinguish between these options by observing the hormone-induced change in NADH when O_2 supply is less compromised than in excised, perifused tissue. In situ measurements of surface fluorescence have been made of hamster brown fat pads with an uncompromised blood supply (Prusiner et al., 1968). These studies of in vivo brown adipose tissue showed that the surface fluorescence due to intracellular NADH decreased when the tissue was exposed to intravenously injected norepinephrine. Similarly, isolated brown fat cells, suspended in media saturated with 95% O_2 exhibited an oxidation of NADH when norepinephrine was added to the suspending media (LaNoue, Koch, and Meditz, 1982). These observations, coupled with the knowledge that anaerobiosis produces a much more rapid depolarization of the cells than ouabain does (Girardier, Seydoux, and Clausen, 1968), suggest that a fall of ATP or an increase in cytosolic NADH may increase Na^+ permeability.

Recent experiments in our laboratory support this concept. Initial rates of ouabain-sensitive $^{86}Rb^+$ uptake and ouabain-sensitive $^{22}Na^+$ fluxes were made in isolated adipocytes (LaNoue, Strzelecki, and Koch, 1984). The adipocytes were suspended in a buffer in which K^+ was replaced with $^{86}Rb^+$. $^{86}Rb^+$ uptake was inhibited 90% by 1 mM ouabain within 30 seconds. Cells were separated from their media by flotation through silicone oil by a modification of the procedure described by Resh, Nemenoff, and Guidotti (1980). The rate of $^{86}Rb^+$ uptake was 0.3 nmoles/min · 10^6 cells, and was unchanged by norepinephrine addition (LaNoue, Strzelecki, and Koch, 1984). Sodium fluxes were ten-fold more rapid than the ^{86}Rb fluxes and, as one would expect, the Na^+ flux was also insensitive to norepinephrine. The rapidity of the Na^+ fluxes and their insensitivity to ouabain suggested the presence of an active Na^+/Na^+ exchange. Evidence for rapid Na^+/Na^+ exchange has been reported in the case of white adipocytes (Resh, Nemenoff, and Guidotti, 1980).

The concentration of intracellular sodium in both the presence and absence of norepinephrine was 8 mM (LaNoue, Koch, and Meditz, 1982). Efflux of sodium from cells preloaded in the presence of O_2 was not altered by anaerobiosis but there was a large influx of labeled sodium in the absence of O_2 or in the presence of both CN^- and iodoacetate. The fact that previous studies showed that $\Delta\Psi_c$ is much more sensitive to iodoacetate than to CN^- (Williams and Matthews, 1974*a*) suggests that there may be a special relationship between ATP generated by glycolysis and the Na^+ fluxes, as has been suggested in other cell types (Fossel and Solomon, 1977).

In very recent, elegant studies of hormone effects on $\Delta\Psi_c$ in excised tissue (Girardier and Schneider-Picard, 1983) it has been demonstrated that norepinephrine addition or nerve stimulation activates the tissue and produces a very fast 5 mvolt depolarization of the cell. This small transient is complete within 10 seconds after nerve stimulation and appears to be mediated by α_1-receptors. The decrease of $\Delta\Psi_c$ is followed by reduction in NADH and subsequently by a larger, sustained depolarization. Evidently the second, but not the first, depolarization is due to anaerobiosis of the excised tissue.

There have been a number of recent studies that have focused on the differentiation of the effects of α- and β-receptors in both white and brown adipocytes (for review see Fain and Garcia-Sainz, 1983). Several laboratories have shown that α_1-agonists can stimulate brown fat respiration, though to a much lesser extent (~20%) than β-agonists (Mohell, Nedergaard, and Cannon, 1983; Sundin and Fain, 1983). The specificity of antagonist effects indicates that α_1-receptors are involved and that these have been clearly identified by measurements of binding of [^{3}H]-prazosin in crude membranes and isolated cells (Mohell, Svartengren, and Cannon, 1983). The K_d of prazosin was 0.4 nM. The α_1-agonists are known to stimulate the phosphoinositol cycle, increasing the level of phosphorylated inositol derivatives in the cytosol, which in turn appear to be involved in the mobilization of intracellular Ca^{2+} stores and possibly in the influx of extracellular Ca^{2+} (Joseph et al., 1984; Streb et al., 1983). The Ca^{2+} ions are second messengers of the α_1 hormones. It has been shown that the α_1-agonists stimulate phosphoinositol turnover in brown fat cells (Fain and Garcia-Sainz, 1983; Strunecka et al., 1981; Garcia-Sainz, Hasler, and Fain, 1980; Mohell, Wallace, and Fain, 1984) and that the α_1-stimulation of respiration does not occur in the absence of external Ca^{2+} (Mohell, Nedergaard, and Cannon, 1983). The β-agonists, in the presence of α_1-antagonists, do not stimulate respiration as efficiently as norepinephrine. Likewise the α_1-antagonists do not stimulate respiration as efficiently as norepinephrine. Thus the α_1- and β-agonists appear to have additive effects. Therefore it is possible that the transient depolarization seen during nerve stimulation is due to small inward Ca^{2+} currents activated by the α_1-agonists.

PROTON CYCLING

The Uncoupling Protein

Initial studies of the isolated brown fat mitochondria indicated that they were uncoupled, since rates of respiration were not stimulated by ADP or by classical uncoupling agents (Drahota, Hanova, and Hahn, 1968; Hohorst and

Raphael, 1968; Smith, Roberts, and Hittelman, 1966). Because it seemed possible that in situ fatty acids were responsible for the uncoupling, efforts were made to deplete the isolated mitochondria of endogenous fatty acids. Endogenous fatty acids could be removed by treating the mitochondria with albumin or by incubating them with carnitine, CoA, and ATP, in order to activate and then oxidize the bound fatty acids (Raphael, Klaas, and Hohorst, 1968; Raphael, Ludolph, and Hohorst, 1969; Hittelman, Lindberg, and Cannon, 1969). These studies revealed that removal of fatty acids alone was not sufficient to recouple the mitochondria. However, addition of adenine or guanine nucleotides to the fatty acid-depleted mitochondria resulted in a preparation of mitochondria whose respiration rates could be stimulated by uncoupling agents. Significantly, ADP was still not as effective as uncouplers in stimulating respiration, due to the low activity of the mitochondrial ATP synthase. Even when stimulated by uncouplers the mitochondrial ATP synthase does not have enough activity, acting in reverse, to hydrolyze the small amount of ATP formed by substrate level phosphorylation via succinyl CoA thiokinase. Thus, in the presence of α-ketoglutarate, brown fat mitochondria will maintain a very high level of ATP/ADP in the external media in both the absence and the presence of uncoupling agents (Nicholls and Bernson, 1977).

Initial experiments suggested that there was a large pool of fatty acids bound to the isolated mitochondria of which only a small fraction was involved in the uncoupling phenomenon (Bulychev et al., 1972; Hittelman, Lindberg, and Cannon, 1969). Later studies (Heaton and Nicholls, 1976) showed that the bound pool of fatty acids, 17 nmoles/mg protein, was reduced to only 1 nmol/mg when recoupling was complete (Heaton and Nicholls, 1976). Isolation of the mitochondria in albumin-containing sucrose media does not lower the fatty acid content enough to recouple the mitochondria by the usual criteria (Hittelman, Lindberg, and Cannon, 1969). Upon standing, fatty acid-depleted mitochondria will also generate enough fatty acids via endogenous lipases to uncouple themselves (Heaton and Nicholls, 1976). It was clear from these studies that the mitochondria required both fatty acid depletion and the presence of adenine of guanine nucleotides for optimal recoupling. The order of effectiveness of the nucleotides was GDP>GTP>ATP>ADP and they were required in mM quantities. The site of action of the nucleotides appears to be on the outer surface of the inner mitochondrial membrane. This conclusion is drawn because the matrix fraction of brown fat mitochondria contains mM levels of endogenous nucleotides (LaNoue, Koch, and Meditz, 1982; Raphael and Wrabitz, 1976), which are not effective in recoupling (Nicholls, 1976*a;* Nicholls and Bernson, 1977), and because inhibitors of mitochondrial adenine nucleotide transport do not alter the effectiveness of the external nucleotides (Nicholls, 1976*a;* Cannon, Nicholls, and Lindberg, 1973).

Further progress in understanding the nature of respiratory control in brown fat was made by Nicholls and co-workers in a well-conceived series of studies that led to the identification of the nucleotide-sensitive uncoupling protein of brown fat (Locke and Nicholls, 1981; Nicholls and Bernson, 1977; Nicholls, 1974*a*, 1974*b*, 1976*a*, 1977; Heaton and Nicholls, 1977). Applying the principles of the chemiosmotic hypothesis, Nicholls measured the ability of the brown fat mitochondria to pump protons, as well as their passive proton permeability (Nicholls, 1974*a*, 1974*b*, 1976*a*, 1977; Heaton and Nicholls, 1977). He observed that they had an undiminished capacity to pump protons but that fatty acid-depleted mitochondria, in the absence of nucleotides, were an order of magnitude more permeable to protons than heart or liver mitochondria were. The proton permeability was estimated by determining the ratio of electron transport to the mitochondrial proton motive force, $\Delta\mu H^+$, by measuring the decay of a proton pulse induced by a burst of respiration, and by observing the rate of mitochondrial swelling in K^+ acetate and valinomycin. Swelling in this situation without a respiratory substrate is dependent on the rate that the proton, brought in with acetate, can efflux passively. In the presence of added purine di- and tri-nucleotides the conductivity was normal. The brown fat mitochondria could maintain a $\Delta\mu H^+$ as high as liver mitochondria ($\sim$200 mvolts) and support an equivalent external phosphorylation potential. It is interesting to note that high external ATP/ADP ratios were generated, in the presence of oxidizable substrates such as α-ketoglutarate, via substrate level phosphorylation catalyzed by succinyl CoA thiokinase. These high external ATP/ADP ratios were maintained even in the presence of uncoupling agents, suggesting a lack of equilibrium between the ATP/ADP ratios and $\Delta\mu H^+$ even at relatively moderate rates of ATP synthesis (Nicholls and Bernson, 1977). This lack of equilibrium is likely due to the low amount of the mitochondrial H^+-ATPase in the brown fat mitochondria.

Further studies showed that GDP ($\sim$0.4 nmoles/mg mitochondrial protein) was bound tightly and specifically to the mitochondria and that GDP binding correlated well with a decrease in proton permeability (Nicholls, 1976*a*). Binding to the mitochondria caused no chemical change in GDP. Brown fat mitochondria also have a high chloride permeability reducible by the same nucleotides that decrease the proton permeability. The decrease in chloride permeability also has the same pH dependence and similar nucleotide specificity as the decrease in proton permeability (Nicholls, 1974*b*). However, the chloride permeability is not as dependent on the level of endogenous free fatty acids as the proton permeability is. In a very recent publication the characteristics of the chloride permeability and the proton permeability have been carefully compared with respect to the nucleotides. The conclusions were drawn that the two conductances, though related, are not due to the same channel and that Cl^- and H^+ are not competitive (Kopecky et al.,

1984), as previously reported. In an effort to determine cause-effect relationships in the physiological regulation, Nicholls has provided additional data concerning the interaction of fatty acids, nucleotides, and the conductance pathway (Nicholls and Locke, 1984). Proton conductance was measured by swelling in potassium acetate plus valinomycin using fatty acid-containing and fatty acid-depleted mitochondria. Proton conductance in the absence of GDP was twice as high in the presence as in the absence of bound fatty acids.

This demonstration that the effect of fatty acids and the absence of nucleotides on proton conductance are not additive is cited as evidence that the fatty acids and nucleotides act on the same target protein (Nicholls and Locke, 1984). Additional support for the concept that the fatty acids act specifically through the uncoupling protein is provided by a comparison of the effect of fatty acids in cold- and warm-adapted guinea pigs (Locke, Rial, and Nicholls, 1982; Rial and Nicholls, 1984). The warm-adapted animals have very low amounts of the uncoupler protein in their mitochondria. Likewise, the isolated mitochondria of the warm-adapted animals are not as sensitive to fatty acid uncoupling as the cold-adapted animals.

The uncoupling protein was originally identified by photoaffinity labeling (Heaton et al., 1978). The ATP-photoaffinity analog 8-azido-ATP, which inhibits [^{3}H]GDP binding in the dark, was used. When irradiated with UV light, the azido derivative labeled two proteins in the brown fat. One of these was identified as the adenine nucleotide carrier; the other was a newly identified protein not found in liver and heart. The unknown protein had a molecular weight of 32,000, a particularly gratifying finding since Ricquier and Kader (1976) had previously identified in rat brown fat mitochondria a protein of similar molecular weight that increased in amount during cold adaptation. The 32,000 D uncoupling protein is present in brown fat mitochondria from all species studied thus far. It was isolated in partially purified form by Ricquier and co-workers (1980) and was later purified to homogeneity by Lin and Klingenberg (1980). Now the protein has been isolated from a variety of species (Ricquier, Lin, and Klingenberg, 1982) and partially characterized (Lin, Hackenberg, and Klingenberg, 1980; Lin and Klingenberg, 1982). The purified protein retains the ability to bind GDP, and successful reconstitution into artificial phospholipid vesicles has been reported recently (Strielman, Schalinske, and Shrago, 1985). In isolation it binds 0.5 nmol GDP/nmol protein which suggests that the functional form is a dimer, and cross-linking studies support this concept.

Lin and Klingenberg (1982) estimate that in hamsters there is three times the amount of uncoupling protein as cytochrome aa_3, on a mol/mol basis. The amount in hamster mitochondria increases during cold adaptation (Ricquier and Kader, 1976; Desautels, Zaror-Behrens, and Himms-Hagen, 1978; Ricquier, Mory, and Hemon, 1979) and in all other rodents during adaptation to a high calorie diet (diet-induced thermogenesis) (Himms-Hagen, Triandafillon, and Gilliam, 1981). Usually it is quantitated by one of two different methods. The two methods involve either measurement of the extent of GDP binding to mitochondria or the quantitation of the 32,000 D protein band by Coomassie blue staining after polyacrylamide gel electrophoresis. The results of these two types of measurements have not always agreed and in some instances the discrepancies are large and of theoretical interest. In genetically obese mice, which exhibit a greatly reduced capacity for thermogenesis, there is very low GDP binding to the mitochondria (Himms-Hagen and Desautels, 1978; Goodbody and Trayhurn, 1982) but normal amounts of the 32,000 D protein band as measured by Coomassie blue staining (Himms-Hagen and Desautels, 1978). This discrepancy has been attributed to inherent lack of precision and specificity in measuring the protein in polyacrylamide gels (Nicholls and Locke, 1984). Recently, reliable immunoprecipitation techniques have been developed by injecting the purified rat protein into ewes (Ricquier et al., 1983*b*). Antibodies to the rat protein prevent GDP binding to the uncoupling protein and cross-react with the protein from all other measured species except the lamb.

The protein has been quantitated by radioimmunassay in the mouse during acclimation to the cold and the results have been compared with GDP binding (Ashwell et al., 1983). The ratio of GDP bound to dimeric uncoupling protein did not vary significantly with cold adaptation, but the ratio was significantly lower than 1, varying randomly from 0.3 to 0.5, raising the possibility that some of the uncoupling protein may exist in an inactive form that is capable of variable, controlled activation.

The availability of the antibody has made studies of the biosynthesis of the protein in cell-free systems possible (Ricquier et al., 1983*a;* Freeman et al., 1983). The newly synthesized protein has a molecular weight of 32,000 before insertion into the mitochondria. Thus, like the adenine nucleotide carrier (Zutzinski, Schleyer, and Neupert, 1983), but unlike some other mitochondrial proteins (Maccechini et al., 1979), it is not synthesized in a higher molecular weight precursor form.

Physiological Regulation of the Uncoupling Protein

To establish the participation of the uncoupling protein and H^+ cycling as a part of the mechanism of hormone action in brown fat, it is important to demonstrate that (1) the protein is a necessary requirement for hormone action; (2) the mitochondria in brown fat cells are coupled in the basal resting state and uncoupled when the cells are exposed to hormone; and (3) there is a feasible means of controlling the activity of the uncoupling protein that is sensitive to the presence of β-receptor agonist.

A drug which could specifically inhibit the uncoupling protein would be the ideal tool for studying the importance of the uncoupling protein in the hormone response, but (unfortunately) such a drug is not yet available. However, it is highly significant that the protein is present in all brown fat mitochondria but in no other organ (Nicholls and Locke, 1984). It is also relevant that the amount of the protein per mg of mitochondria and, more dramatically, per animal increases when the capacity of the animal for thermogenesis increases due to cold exposure, diet, and so forth (Cannon, Nedergaard, and Sudin, 1981; Desautels, Zaror-Behrens, and Himms-Hagen, 1978; Ricquier, Mory, and Hemon, 1979; Himms-Hagen, Triandafillon, and Gilliam, 1981; Goodbody and Trayhurn, 1982; Ashwell et al., 1983).

This correlation of the amount of the 32,000 D protein with the thermogenic capacity of the animal implies that the protein is rate-limiting for thermogenesis. This implication is puzzling in view of what is known about its kinetic parameters. The K_d of the protein for purine nucleotides is in the μM range (Nicholls, 1974*a;* Lin and Klingenberg, 1982). Values in this range are obtained by using the isolated protein (Lin and Klingenberg, 1982) and also from binding studies with isolated brown fat mitochondria in the presence of carboxyatractyloside and from studies of the effect of ATP on mitochondrial swelling both in chloride salts and in potassium acetate plus valinomycin (Nicholls, 1974*a;* Heaton and Nicholls, 1977). Yet when the effects of ATP and GTP are measured on respiration or on the mitochondrial membrane electrical gradient (Horwitz and Easton, 1975; Locke and Nicholls, 1981; LaNoue, Koch, and Meditz, 1982; Heaton and Nicholls, 1976, 1977; Locke, Rial, and Nicholls, 1982), the half maximal effective concentration of the nucleotides in the presence of physiological concentration of Mg^{2+} levels is about 1 mM. In the absence of Mg^{2+}, the half maximal effective concentration of GDP on respiration is 100 μM, more than an order of magnitude greater than the K_d (Nicholls, 1974*a*). The lack of agreement between the binding constant and the amount of nucleotides needed for coupling has been attributed to the fact that there is a large excess of uncoupling protein in the membrane (Nicholls, 1979). Evidently the vast majority of the channels (not just 50%) must be closed before a significant $\Delta\Psi_m$ can develop. This excess capacity of the protein does not appear to be consistent with the concept that it is the limiting factor in thermogenic capacity.

Although definitive proof is lacking, the majority of the available data strongly support the concept that the mitochondria are coupled in the basal state and uncoupled in the hormone-stimulated state. Using isolated, intact cells other than brown adipocytes, the ratio of dinitrophenol or carbonyl cyanide p-trifluoromethoxyphenylhydrazone (FCCP) uncoupled respiration to oligomycin-inhibited respiration is frequently used to test the extent of in situ coupling (LaNoue, Strzelecki, and Finch, 1984; Watts, Koch, and LaNoue, 1980). Oligomycin inhibits the mitochondrial ATP synthase so that it can no longer respond to the presence of ADP and phosphate in the external media (Nicholls, 1982). It therefore prevents both the flow of protons into the mitochondrial matrix through the ATP synthase and the resultant stimulation of respiration. Oligomycin-inhibited respiration is analogous to the resting (state 4) rate of mitochondrial respiration observed in the absence of phosphate acceptors. Addition of an uncoupling agent such as FCCP bypasses this inhibition by providing an alternate route for protons into the mitochondria.

Early attempts to use uncoupling agents and oligomycin to assess the degree of coupling in brown fat were unsuccessful because both reagents inhibited the respiratory response induced by norepinephrine (Nedergaard and Lindberg, 1982; Prusiner, Cannon, and Lindberg, 1968; Williamson, 1970; Pettersson, 1977). Moreover, the inhibition due to oligomycin was not reversed by uncouplers. It soon became apparent that the usefulness of these two reagents was compromised by the fact that they lower tissue ATP levels. The adipocytes are almost entirely dependent on fatty acids as respiratory fuel (Bukowiecki et al., 1981) and a certain minimal level of ATP in the cytosol is required to activate fatty acids. Thus, the tissue becomes substrate-limited when ATP levels fall.

Other data do suggest however that the in situ mitochondria go from a coupled to an uncoupled state when cells are exposed to β-agonists. The addition of an uncoupling agent will stimulate a twofold increase in the basal rate of respiration (Prusiner, Cannon, and Lindberg, 1968; LaNoue, Koch, and Meditz, 1982; Williamson, 1970). Treatment with hormone lowers tissue ATP levels (Pettersson and Vallin, 1976; LaNoue, Koch, and Meditz, 1982; Williamson, 1970; Knight, 1974) and causes an oxidation of both NADH (Prusiner et al., 1968; LaNoue, Koch, and Meditz, 1982) and cytochrome b (Prusiner, 1970). This effect implies a stimulation of the electron transport chain, which in view of the limited kinetic capacity of the mitochondrial ATP synthase is most logically attributed to a decrease in $\Delta\Psi_m$, the mitochondrial electrical potential, mediated by the uncoupling protein. Likewise, decreased ATP and NADH levels are not consistent with the concept that the mitochondria are controlled by substrate.

In an effort to circumvent the limitation of substrate supply in the presence of oligomycin and uncouplers, LaNoue, Koch, and Meditz (1982) added additional substrates to isolated adipocytes. Succinate stimulated basal respiration of the cells 3-fold and addition of durohydroquinone stimulated respiration 10- to 20-fold. This dramatic stimulation by the unphysiological quinone was not dependent on the presence of norepinephrine. Durohydroquinone-stimulated respiration was only slightly stimulated by uncoupling agents and was not inhibited by oligomycin. Durohydroquinone is an analogue of ubiquinone and interacts with the electron transport chain at the level of coenzyme Q (Lawford and Garland, 1972; Ruzicka and Crane, 1971). Since coenzyme Q is on the oxygen side of coupling site I of ATP synthesis, energy in the form of $\Delta\mu H^+$ is required for both durohydroquinone and succinate to reduce mitochondrial NADH (Klingenberg, 1964). Measurements of cellular NADH before and after addition of succinate or durohydroquinone indicated that succinate doubled and durohydroquinone tripled the NADH levels. However, in the presence of norepinephrine NADH levels remained very low and were not increased by addition of succinate or the quinone. Reverse electron transport through a coupling site of ATP synthesis requires an electrochemical gradient of protons across the mitochondrial membrane (Nicholls, 1982). The data show that reverse electron transport occurs in the basal but not the hormone-stimulated state, suggesting a fairly large difference in $\Delta\mu H^+$ between the two states. To confirm this difference, the electrical potential gradients across the mitochondrial membranes have been measured in brown adipocytes using a method proved useful in synaptosomes (Scott and Nicholls, 1980), hepatocytes (Hoek, Nicholls, and Williamson, 1980), lymphocytes (Brand and Felber, 1984; Arslan et al., 1984), and white adipocytes (Davis and Martin, 1982). To carry out the measurement, a lipophilic cation such as tetraphenylphosphonium (TPP^+) or triphenylmethyl phosphonium ($TPMP^+$) is incubated with the cells and the accumulation of the cation is measured. When the accumulation is appropriately corrected for distribution coefficients and accumulation due to the electrical potential of the cell membrane, it is possible to calculate the mitochondrial membrane potential. The method is most thoroughly described by Brand and Felber for use with lymphocytes (Brand and Felber, 1984). An improved method of correcting for binding of the lipophilic cation to mitochondrial membranes has also been published recently (Rottenberg, 1984). When the method was employed by LaNoue and Strzelecki (1983) to measure the $\Delta\Psi_m$ of brown adipocytes in the basal state, $\Delta\Psi_m$ was found to be surprisingly low, about 116 mvolts, while the corresponding measurement carried out in hepatocytes at the same time gave the more normal value of 140 mvolts (LaNoue et al., 1986; LaNoue and Strzelecki, 1983).

Addition of norepinephrine to brown adipocytes lowered the $\Delta\Psi_m$ by about 60 to 56.6 mvolts. Figure 1-1 is an illustration of the data obtained in both the presence and absence of norepinephrine. To calculate $\Delta\Psi_m$, one must correct for cation accumulation due to $\Delta\Psi_c$, the cytosolic potential. To make the correction, the potential across the cell membrane was assumed to be constant and equal to 42 mvolts, the value obtained from in situ microelectrode studies (Horwitz, Horowitz, and Smith, 1969; Flaim, Horwitz, and Horowitz, 1977). The assumption that $\Delta\Psi_c$ is the same in both the basal and hormone-stimulated states seems justified on the basis that cell membrane depolarization measured in the presence of an adequate O_2 supply is a transient phenomenon, lasting only a few minutes or seconds, whereas the measurement using lipophilic cations requires more than an hour to complete. The data obtained and the method of calculation are described in detail by LaNoue and co-workers (1986).

These results may appear to be at variance with the observations on isolated brown fat mitochondria suggesting that the recoupled mitochondria can attain normal high values of $\Delta\Psi_m$ and $\Delta\mu H^+$ in the range of 150 to 200 mvolts, but they are in agreement with similar measurements made by Nicholls on brown fat adipocytes in the basal state (Nicholls and Locke, 1984; Raphael and Nicholls, 1984) when $TPMP^+$ was used as the indicator. However, using a different method to measure $\Delta\Psi_m$ in the cell, Nicholls obtained much higher values for $\Delta\Psi_m$ in situ in both the basal and hormone stimulated states (Davis and Martin, 1982). To obtain the higher values he employed the fluorescent dye, 2- (dimethylaminostyryl)-1-methylpyridinium, which accumulates in lipid bilayer vesicles according to their electrical potential gradient. A 400-fold accumulation of the dye in the cells was

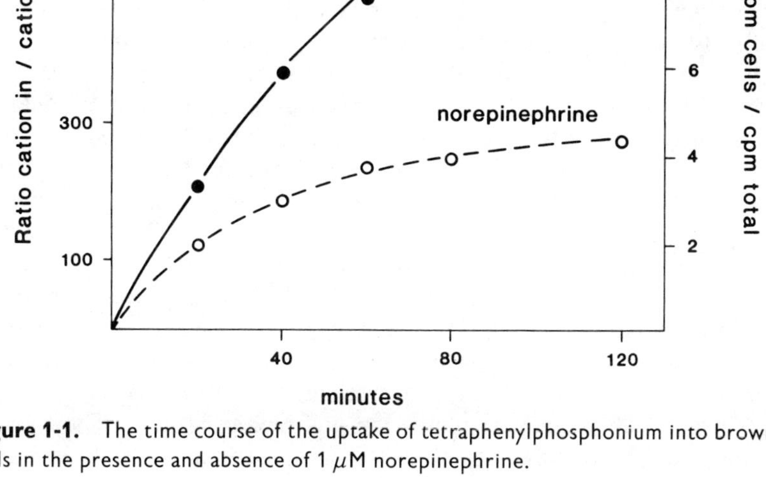

Figure 1-1. The time course of the uptake of tetraphenylphosphonium into brown fat cells in the presence and absence of 1 μM norepinephrine.

reported. It appears, however, that the data were not appropriately corrected for $\Delta\Psi_c$ or distribution coefficients. In the absence of any electrical potential gradient, 4-fold accumulation was observed, which provides an indication of the magnitude of the passive distribution coefficient. If one corrects for the apparent distribution coefficient a ratio of 100-fold for the free dye accumulation is obtained. Since the $\Delta\Psi_c$ is 42 mvolts (Horwitz, Horowitz, and Smith, 1969; Flaim, Horwitz, and Horowitz, 1977) and the mitochondrial matrix space is 40% of the cell volume (LaNoue, Koch, and Meditz, 1982), the ratio of dye in cytosol to dye in the mitochondria would appear to be only about 14, again a very low value compared to other cell types. It therefore appears to the present reviewer that the $\Delta\Psi_m$ is usually low in the basal state but that it virtually disappears in the hormone-stimulated state. All of the available data support the concept that proton cycling via the uncoupling protein is important in brown fat thermogenesis. However, as mentioned above, to establish the participation of the uncoupling protein in the mechanism of hormone action, one must demonstrate a feasible means of controlling its activity by the hormone.

The mechanism of physiological control of the activity of the uncoupling protein has been a matter of conjecture for some time. Recent studies appear consistent with the concept that control is mediated by free fatty acids inside the cell (Locke and Nicholls, 1981; Locke, Rial, and Nicholls, 1982; Locke et al., 1982), the levels of which are regulated by the hormone-sensitive lipase. Nevertheless, other interpretations of the data, which suggest that cytosolic ATP may be the direct physiologic regulator, are possible and the data supporting this alternate view will also be discussed.

Evidence in favor of control by fatty acids comes from several different types of experiments. Perhaps the most important of these is the observation that fatty acids added outside the brown fat cells will stimulate respiration as effectively as norepinephrine will (Prusiner, Cannon, and Lindberg, 1968; Williamson, 1970; Bukowiecki et al., 1981). The concentrations of fatty acids and the ratio of fatty acid to serum albumin required for stimulating respiration are higher than those present in the circulating blood. This fact, of course, must be true because thermogenesis is not stimulated physiologically by high levels of fatty acids in the blood but by norepinephrine released at the sympathetic nerve endings (Himms-Hagen, 1976). Preliminary data (unpublished) from our laboratory suggest that the same ratio of fatty acid to albumin which stimulates respiration also blocks reverse electron transport from durohydroquinone to NADH, implying that the same concentrations of fatty acids stimulate respiration and uncouple it simultaneously. Therefore it appears that either fatty acids or a cytosolic metabolite produced by the catabolism of fatty acids regulates the uncoupling protein. The obvious candidates are fatty acids themselves, fatty acyl carnitine and fatty acyl CoA.

Since endogenous fatty acids must be removed in order to recouple the isolated mitochondria and since very low levels of external fatty acids will uncouple the fatty acid-depleted mitochondria, it was not unreasonable to propose that free fatty acids in the cell regulate the activity of the uncoupling protein. Long chain acyl carnitine is not a potent uncoupling agent in brown fat (Locke and Nicholls, 1981; Locke et al., 1982). On the other hand, it has been proposed that long chain fatty acyl CoA is the physiological effector, and Cannon, Sundin, and Romert (1977) have shown that long chain acyl CoA is able to reverse the inhibition of chloride conductance induced by purine nucleotide addition. However, the amount of long chain acyl CoA required for this effect is greater than 10 nmoles/mg mitochondrial protein. LaNoue, Koch, and Meditz (1982) found that the total cellular content of long chain acyl CoA increased from 0.5 to 0.7 nmoles/mg mitochondrial protein on addition of norepinephrine to intact isolated adipocytes. Moreover, compartmentation studies suggested that no more than 0.1 nmole/mg of mitochondria could have been on the cytosolic side of the membrane in either condition.

Interesting and important experiments by Locke and Nicholls (1981; Locke et al., 1982) with isolated recoupled brown fat mitochondria also gave results consistent with the concept that fatty acids are the direct physiological regulators of the uncoupling protein. Experiments were carried out with isolated mitochondria suspended in a media containing pyruvate, bovine serum albumin, Mg^{2+}, 2 mM ATP, CoA, carnitine, and TPP^+. After a seven-minute preincubation period during which both O_2 consumption and $\Delta\Psi_m$ were monitored using respectively an O_2 electrode and a TPP^+-sensitive electrode, fatty acids were infused at a rate of 13 nmoles/min/mg of protein with continuous monitoring of O_2 consumption and $\Delta\Psi_m$. At 30°C, fatty acids with no detectable threshold concentration increased respiration from 43 to 111 natoms O/min · mg mitochondrial protein and $\Delta\Psi_m$ decreased from 225 to 215 mvolts. The experiments demonstrate that low concentrations of fatty acids can stimulate respiration of recoupled brown fat mitochondria in the presence of mM levels of ATP.

Other recent experiments with isolated recoupled brown fat mitochondria suggest that the $\Delta\Psi_m$ varies only slightly under physiological conditions. Nedergaard (1983), for example, has reported that $\Delta\Psi_m$ decreases from 150 mvolts in the presence of 1 mM GDP to 135 mvolts at 1 μM GDP in the presence of 0.2% fatty-acid-free bovine serum albumin. The measurements were carried out using saffranine dye as an indicator of $\Delta\Psi_m$. Trayhurn and Fraser (1983), using $TPMP^+$ gradients to measure $\Delta\Psi_m$ in isolated mitochondria from brown fat of lean controls and genetically obese mice, found that the addition of GDP increased $\Delta\Psi_m$ from 94 to 150 mvolts in controls while the same concentration of GDP increased $\Delta\Psi_m$ from 115 to 135 mvolts in mitochondria from brown fat of mature obese mice. The conclusion was drawn that the uncoupling protein had lower activity in the obese mice than in the controls.

An obvious problem is that the measurements of $\Delta\Psi_m$ in isolated mitochondria appear to be at significant variance with measurements of $\Delta\Psi_m$ in the cell. The rates of mitochondrial respiration are also very different in the cell than they are after the mitochondria are isolated. The isolated mitochondria respire at far lower rates than those inside the cell. Mitochondrial content of cells can be determined by measuring the amount of mitochondrial marker enzymes in cells or tissue sonicates. Data from a number of sources are in agreement and suggest that about ¼ to ½ of the cellular protein is mitochondrial and that there is about 1.5 mg of total protein per 10^6 cells (LaNoue, Koch, and Meditz, 1982; Raphael and Wrabitz, 1976).

Since the rate of norepinephrine-stimulated respiration at 37°C in isolated brown adipocytes is about 800 natoms O/min · mg the rate of respiration of mitochondria in brown fat cells should be 1,400 natoms O min/mg when measured at 37°C. Brown fat mitochondrial respiration is usually reported at 20°C to 25°C and values vary from about 100 to 200 natoms O/min · mg mito, using palmitoyl carnitine, succinate, α-ketoglutarate, or α-glycerophosphate. Assuming that respiration increases twofold for every 10°C, the rates are measured at 37°C would be from 300 to 500 natoms/min · mg mitochondrial protein, still far below that suggested by the cellular data. In our hands, mitochondria prepared by the standard procedures and measured at 37°C respire at the rate of ~300 natoms/min · mg (LaNoue, Koch, and Meditz, 1982). If, however, the mitochondria are isolated in media containing 5 mM ATP, the osmotic collapse (Nicholls, Gray, and Lindberg, 1972), caused by the unusual ion permeabilities of the brown fat mitochondria, is prevented. Prepared in the usual way, brown fat mitochondria must be incubated in media of very low ionic strength (e.g., 175 milliosmoles) (Nedergaard, 1983) in order to obtain even the low O_2 consumption normally reported. On the other hand, at 37°C, the mitochondria prepared in 5 mM ATP respire at the rate of 1200 to 1500 natoms O/min · mg mito with palmityol carnitine as substrate in isotonic KC1 media (LaNoue, Koch, and Meditz, 1982). Other physiological substrates such as pyruvate and succinate are oxidized relatively slowly (i.e., 200-300 natoms O/min · mg mito) in agreement with the behavior of in situ mitochondria.

The respiration rates of mitochondria prepared in the presence of ATP are inhibited by the addition of ATP and appear to be coupled without the recoupling procedures (LaNoue, Koch, and Meditz, 1982). Even though they contain 20 nmoles fatty acid/mg mitochondrial protein (LaNoue et al., 1986), respiration in the presence of palmitoyl carnitine and 5 mM ATP is stimulated threefold to fivefold by the addition of uncoupling agents. Measurement of $\Delta\Psi_m$ shows that the proton conductance of the membrane is an order of magnitude higher, even in the presence of 5 mM ATP, than the conductance of normal liver mitochondria or of brown fat mitochondria depleted of fatty acids. The $\Delta\Psi_m$ of the mitochondria isolated in the presence of ATP but retaining the endogenous fatty acids was 41.6 mvolts and addition of ATP increased $\Delta\Psi_m$ to 116 mvolts. On the other hand, when these mitochondria were depleted of fatty acids, the $\Delta\Psi_m$ was 119 mvolts in the absence of ATP and only 123 mvolts in the presence of 5 mM ATP (LaNoue et al., 1986).

Despite the very high proton conductance, control by the uncoupling protein occurs, because the very high capacity for electron transport (and proton pumping) competes effectively with the high degree of proton permeability. However, when ATP levels are low the proton conductance pathway is stimulated relative to respiration and $\Delta\Psi_m$ falls.

Therefore it seems likely that even in the absence of hormone stimulation the in situ mitochondria contain bound fatty acids in amounts previously considered to be uncoupling. The recent experiments of Locke and co-workers (1982), in which it was shown that very low quantities of free fatty acids uncouple brown fat mitochondria, imply that the concentration of fatty acids required for activation of fatty acyl CoA synthase may be higher than that needed for the increase in proton conductivity. Measurements of intracellular fatty acids in brown adipocytes made several years ago by Bieber, Pettersson, and Lindberg (1975) indicated that the concentration of fatty acids in the basal and hormone-stimulated state is 20 mM. Although it is not possible to know the ratio of bound to free fatty acids in the cell, it seems likely that the levels of cytosolic free fatty acids even in the basal state are sufficient to produce a fairly high conductivity to protons in the intracellular mitochondria.

However, it appears to the present reviewer that fatty acids themselves do not directly regulate the change in $\Delta\Psi_m$ observed in the intact cells after hormone addition. A more likely possibility is that the uncoupling protein is regulated by ATP levels. Although the measured changes in cellular ATP are not large enough to account for the observed changes in coupling, cytosolic ATP levels may change much more dramatically than total cellular ATP. Earlier claims that cellular ATP concentrations are too high to control the protein (Nicholls, 1976*b;* 1979; Raphael and Wrabitz, 1976) are incorrect since it has been shown by Locke and Nicholls (1981), Heaton and Nicholls (1977), and LaNoue, Koch, and Meditz (1982) that mM quantities of ATP are required in order to lower proton conductance of isolated brown fat mitochondria to levels which inhibit respiration.

Recent studies of adenine nucleotide compartmentation in brown fat cells indicate that the majority of the cellular nucleotides are located within the mitochondrial matrix (LaNoue, Koch, and Meditz, 1982). When the mitochondria are coupled and have a significant $\Delta\Psi_m$, the electrogenic nature of the mitochondrial adenine nucleotide carrier, which catalyzes the exchange of ATP^{4-} for ADP^{3-} (LaNoue, Mizani, and Klingenberg, 1978; Klingenberg and Rottenberg, 1977), makes the ratio of ATP/ADP higher outside the mitochondria than inside. In isolated brown fat mitochondria, the ratio of the external ATP/ADP ratio to the internal ratio is approximately 11 in the coupled state but close to 1 when the mitochondria are uncoupled. Nearly identical data on this point were obtained by LaNoue, Koch, and Meditz in 1982 and by Raphael and Wrabitz in 1976 using more conventionally prepared mitochondria. This means that nearly all of the cellular ATP is cytosolic in the resting non-thermogenic condition, while a large fraction of the measured cellular ATP may be in the mitochondria in the hormone-stimulated state. Since only cytosolic and not mitochondrial nucleotides may regulate the uncoupling protein, it seems that ATP control of the protein is feasible. This scheme for control has been criticized on the grounds that it explains only the steady-state situation and cannot account for the transitions from resting to stimulated state and back (Nicholls and Locke, 1984). A means of hormonal regulation of ATP utilization is required in the proposed scheme, and although no such hormonally sensitive ATPase has been identified, regulated processes that utilize ATP have not been adequately studied.

A reasonable scheme for the sequence of events leading to the stimulation of respiration in brown fat should include an initial stimulation of a cytosolic ATP utilizing process, which would slightly lower cytosolic levels of ATP. Calculations of the concentration of cytosolic ATP reveal that it is in the region of the Km for inhibition of respiration (LaNoue, Koch, and Meditz, 1982). Thus any small decrease in cytosolic ATP could cause a corresponding decrease in $\Delta\Psi_m$ that would allow some cytosolic ATP to enter the mitochondrial matrix. The resultant decrease in cytosolic ATP would amplify the initial decrease until full uncoupling was achieved. Reversal from the hormone-stimulated state to the basal state could be achieved by reversing this sequence of events. Thus the hypothetical ATP-utilizing process should diminish when hormone levels decrease, causing an increase in cytosolic ATP, an increase in $\Delta\Psi_m$, an amplification of the initial increase in ATP, and so on. Independent of the exact mechanism of regulation of the uncoupling protein, it is necessary to postulate a hormone-stimulated ATPase to balance the excess production of ATP by succinyl CoA thiokinase in the thermogenic phase. Complete oxidation of a long chain fatty acid such as palmitate with consumption of 700 natom of O/min $\cdot$ 10^6 cells involves the production of 117 nmoles of ATP/min $\cdot$ 10^6 cells or about 260 nmoles of ATP/min $\cdot$ mg of mitochondrial protein via substrate level phosphorylation by succinyl CoA thiokinase. Since the V_{max} of the mitochondrial ATPase in the uncoupled state is lower than 50 nmoles/min · mg mitochondrial protein (Cannon and Vogel, 1977; Bulychev et al., 1972), another means of ATP breakdown with very high capacity is required, to be regulated in concert with fatty acid oxidation.

There is a significant delay between the addition of hormone to brown adipocytes and the stimulation of respiration. The delay is about one minute at 37°C (Knight, 1974) and two minutes at 28°C (Williamson, 1970). If our scheme for the control of the uncoupling protein is correct, it implies that the fall in cellular ATP would occur at the same time, or prior to, the increase in respiration. The time course of the decrease in cellular ATP has been measured by two laboratories. In one case, when measurements were made at 37°C, the change in ATP appeared to occur at the same time as the increase in respiration (Knight, 1974). When the measurements were carried out at 28°C, allowing better time resolution (Williamson, 1970), the decrease in ATP preceded the increase in respiration.

SUBSTRATE CYCLING

Our original scheme for control of the uncoupling protein by cytosolic ATP included the suggestion that stimulation of Na^+, K^+-ATPase caused the initial, physiologically important decrease in cytosolic ATP (LaNoue, Koch, and Meditz, 1982). However, in situ measurements of ouabain-sensitive ion fluxes showed that there was no change in flux through the pump (LaNoue, Strzelecki, and Koch, 1984). Since fatty acids added to cell suspensions stimulate respiration (Bukowiecki et al., 1981), lower ATP (Williamson, 1970), and uncouple the mitochondria, it appears reasonable to assume that the required ATP-utilizing system is part of a futile substrate cycle involving fatty acids. In fact, two such cycles have been identified in brown fat cells (Newsholme and Crabtree, 1976; Brooks, Arch, and Newsholme, 1983) and are illustrated in Figure 1-2. In this figure the double vertical line represents the mitochondrial inner membrane and the dotted arrows are meant to indicate multistep enzymatic pathways.

Both of these futile cycles were discussed by Newsholme as potentially important in brown fat thermogenesis in review of substrate cycling (Newsholme and Crabtree, 1976). At that time few reliable estimates of the actual rate of cycling and therefore of ATP hydrolysis were available. Since then Newsholme and co-workers (Brooks, Arch, and Newsholme, 1983) have quantitated the triacyl glycerol fatty acid cycle in brown fat cells of mice by injecting ^3H_2O into mice one hour before sacrifice. The brown fat is excised at sacrifice; the triglyceride is extracted and the [^{3}H] label is measured separately in the glycerol and fatty acid moieties. Cycling is calculated as three times the apparent rate of glycerol synthesis minus the rate of fatty acid synthesis. Synthesis is calculated from the amount of [^{3}H] incorporated into neutral fat divided by the specific activity of ^3H_2O in the blood plasma at the time of sacrifice. Data are reported as μmoles/hr/g fat-free dry tissue. Assuming that all the fat-free dry tissue is protein and that there are 1.5 mg protein/10^6 cells, it is possible to estimate the rate of ATP hydrolysis due to this futile

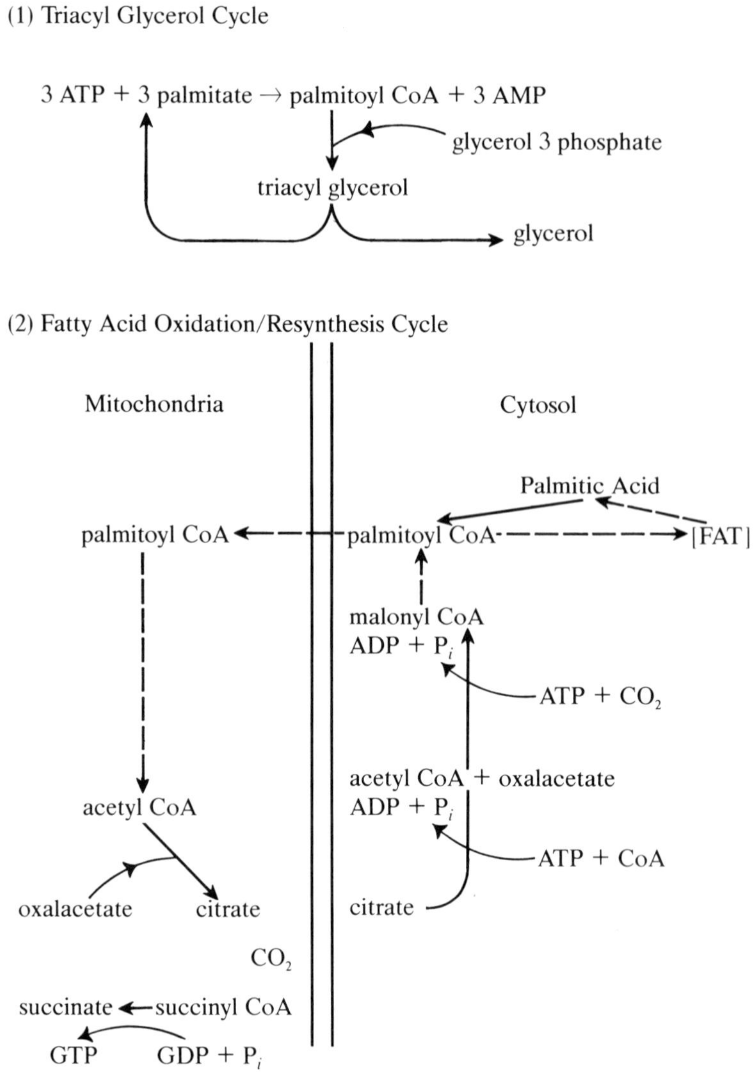

Figure 1-2. Triacyl glycerol cycle and fatty acid oxidation/resynthesis cycle.

cycle and to compare it with the increased rate of ATP synthesis that occurs during hormone stimulation. Data are provided on the effects of a β-agonist, fenoterol, and acute cold exposure. The data suggest that the cycling uses 12.4 nmoles ATP/min · 10^6 cells in the control state, 35.2 after fenoterol injection, and 25.8 during acute cold exposure. Comparing this to the rates required for extra ATP hydrolysis in the hamster during thermogenesis (117 nmoles/min · 10^6 cells) it would appear that the cycle could provide at least a significant fraction of the needed ATP hydrolysis.

The fatty acid oxidation/resynthesis futile cycle is extremely wasteful energetically and could account for the remainder of the ATP utilization. Since 2 moles of ATP are consumed for each mole of acetyl CoA cycled, and since cycling bypasses ATP formation via succinyl CoA thiokinase, 3 moles of ATP per mole of acetyl groups cycled are consumed. The cycle is potentially more wasteful for coupled mitochondria than for uncoupled mitochondria. Fatty acid futile cycling is prevented in normally coupled tissues with high capacity for fatty acid synthesis, such as liver and white adipocytes, by a feedback inhibitor mechanism involving malonyl CoA.

The first committed metabolite in fatty acid synthesis is malonyl CoA. The concentration of malonyl CoA increases with increased rates of formation of cytosolic citrate. High levels of cellular free fatty acids could increase cytosolic citrate formation. Nevertheless, cycling is normally blocked because malonyl CoA is a potent inhibitor of palmitoyl carnitine CoA transferase I (Brooks, Arch, and Newsholme, 1983). Therefore, citrate formation from fatty acyl CoA is subject to feedback inhibition via malonyl CoA, whereas citrate formation from pyruvate, short, and medium chain carboxylic acids is not subject to inhibition by malonyl CoA, allowing fatty acid synthesis from carbohydrates to occur but preventing fatty acid futile cycling. The increased ketone body formation in the livers, which occurs after glucagon administration, involves an increase of flux through palmitoyl carnitine CoA transferase I, while fatty acid synthesis is inhibited (McGarry and Foster, 1979). Malonyl CoA levels are reduced by glucagon (McGarry and Foster, 1979, 1980). One reason for the decrease in malonyl CoA is the diversion of pyruvate and lactate carbon from oxidation to glucose formation; another, perhaps more important reason for the decrease, is that the glucagon-induced increase in cyclic AMP activates a protein kinase, which phosphorylates acetyl CoA carboxylase, the enzyme producing malonyl CoA. Phosphorylation inhibits the enzyme via a decrease in V_{max} and an increase in the Km for citrate (Holland, Witter, and Hardie, 1984). The phosphorylation of the enzyme correlates well with the decrease in fatty acid synthesis (Holland, Witter, and Hardie, 1984). Adrenalin inhibits fatty acid synthesis in white adipocytes by a similar mechanism involving a cyclic AMP dependent protein kinase and phosphorylation of acetyl CoA carboxylase (Witter, Tipper, and Bacon, 1983).

Brown adipocytes have a very high capacity for fatty acid synthesis. The enzymes of fatty acid synthesis, ATP citrate lyase, acetyl CoA carboxylase, and fatty acid synthase have higher activities in brown fat than in liver (Gnoni et al., 1983). Original estimates (Masoro, 1963; Knight and Myant, 1970, 1971) of rates of synthesis using ^{14}C-glucose and ^{14}C-acetate incorporation were underestimated because of lack of adequate oxygenation of tissue fragments and because of lack of precision in estimating isotopic dilution by endogenous substrates (for discussion see McCormack, 1982). Recently, more accurate techniques have involved measurements of the rate of incorporation of [^{3}H] from ^3H_2O into the triglycerides of brown fat of rats (Gnoni et al., 1983; McCormack, 1982; Agius et al., 1981; Agius and Williamson, 1980; McCormack and Denton, 1977; Trayhurn, 1979; Sugden et al., 1982; Marchall, Watts, and Sugden, 1983; Sugden et al., 1983; Agius, Watts, and Sugden, 1983; Tokuyama and Kuda, 1983; Wright and Agius, 1983; Agius et al., 1983; Klein and Williamson, 1982), mice (Path, Salmon, and Hems, 1979; Van den Brandt and Trayhurn, 1981; Trayhurn, 1981), and hamsters (Trayhurn, 1980). As in the studies of the triacylglycerol cycle (Brooks, Arch, and Newsholme, 1983), tritiated water is injected into the animals about one hour before sacrifice, plasma ^3H_2O specific activity is determined at the time of sacrifice, and neutral fat is analyzed for tritium-labeled fatty acids. Isotope effects have been ignored in these studies so that literature values are minimal estimates of flux. An assumption is made that all fatty acyl groups synthesized are incorporated into neutral fat at a constant rate during the one hour in vivo. Data obtained by separate workers under various conditions are shown in Table 1-1.

In each case synthesis was measured by ^{3}H incorporation into fatty acid moieties of neutral fat after an injection of ^3H_2O about one hour prior to sacrifice. In the original references the data are reported as μg atoms ^{3}H incorporated/hr/gm wet wt. They are converted in Table 1-1 to nmoles C_2 units/min · 10^6 cells, assuming 1.75 μg atoms H/C_2 unit and from measuring values of 126 mg protein/gm wet weight and 1.5 mg protein/10^6 cells. The values shown are minimum values for fatty acid synthesis since no account is taken of isotope effects or of those fatty acid molecules synthesized but either exported from the cell or oxidized, rather than incorporated into triacyl glycerol. The temperature of adaptation is the temperature of the animals environment three to four weeks prior to the experiment and experimental temperature is the temperature of the animals' environment during the time of ^{3}H incorporation into the triacyl glycerol.

It is apparent that the rate of cycling is higher in the thermogenic state than in the resting state. The rates of fatty acid synthesis reported in the data of Table 1-1 are sufficient in the more highly thermogenic states to balance the rates of ATP generation by substrate level phosphorylation. However, some of the data of Table 1-1 and other experimental findings suggest that

Table 1-1. Rates of in vivo fatty acid synthesis in brown fat tissue in different metabolic states

Species	*Temperature of Adaptation °C*	*Experimental Temperature °C*	*Rates of Fatty Acid Synthesis nmoles C^2/ min · 10^6 Cells*	*Drugs or Special Conditions*	*References*
Rat	Room temp.	Room temp.	6.30	None	Gnoni et al., 1983
			3.16	None	Path, Salmon, and Hems, 1979
			1.82	None	Trayhurn, 1979
			3.66	None	Sugden et al., 1982
			2.35		
			AV 3.46		
	Room temp.	Room temp.	18.5	+Glucose	Agius and Williamson, 1980
			11.2		Sugden et al., 1982
			AV 14.9		
	Room temp.	Room temp.	16.2		Agius and Williamson, 1980
			21.9	Insulin	Sugden et al., 1982
			AV 19.1		
			2.6	Norepinephrine	McGarry and Foster, 1980
	4°	28°	7.07	None	McCormack and Denton, 1977
			4.84	None	Trayhurn, 1979
			11.5	None	Sugden et al., 1983
			15.8	None	Wright and Agius, 1983
			AV 9.80		
	28°	4°	1.75	None	Trayhurn, 1979
	4°	4°	24.9	None	Trayhurn, 1979
Mouse	Room temp.	Room temp.	5.57		Path, Salmon, and Hems, 1979

(continued)

Table 1-1. *(continued)*

Species	*Temperature of Adaptation °C*	*Experimental Temperature °C*	*Rates of Fatty Acid Synthesis nmoles C^2/ min · 10^6 Cells*	*Drugs or Special Conditions*	*References*
	Room temp.	4° (1 hour)	21.1		Path, Salmon, and Hems, 1979
	Room temp.	4° (24 hours)	73.9		Path, Salmon, and Hems, 1979
	4°	33°	3.19		Trayhurn, 1981
	4°	22°	12.1		Trayhurn, 1981
	4°	4°	42.3		Trayhurn, 1981
	4°	4°	40.2		Van den Brandt and Trayhurn, 1981
	21°	21°	0.6	Suckling mouse	Trayhurn, 1981
Hamster	30°	30°	1.27		Trayhurn, 1980
	30°	4°	0.47		Trayhurn, 1980
	4°	30°	11.5		Trayhurn, 1980
	4°	4°	10.1		Trayhurn, 1980
	4°	30°	5.14	Triton WR 1339 (inhibits lipoprotein lipase)	Trayhurn, 1980

the correlation between the thermogenic state and the rates of fatty acid synthesis is not good. For example, the suckling mouse is known to be highly thermogenic even at 21°C and rates of fatty acid synthesis appear slow. Also, glucose and insulin appear to stimulate fatty acid synthesis as much as exposure to cold does, perhaps due to covalent activation of pyruvate dehydrogenase and acetyl CoA carboxylase. High calorie, high fat diets inhibit ^{3}H-incorporation into neutral fat, yet it is known that the same high calorie diets cause proliferation and activation of brown adipose tissue (diet-induced thermogenesis). Moreover, injection of norepinephrine did not appear to stimulate the incorporation of $^{3}H_2O$ into triglycerides. The lack of correlation between activation of the tissue and rates of fatty acid synthesis may, however, be more apparent than real because simultaneous stimulation of lipolysis and lipogenesis would tend to mask the rate of lipogenesis, since the fatty acid molecules synthesized would be diluted by the fatty acid molecules generated by lipolysis and many of the de novo synthesized molecules would be oxidized by the mitochondria or exported from the cell rather than be incorporated into neutral fat. Further experiments are required to demonstrate the extent of the proposed underestimates and to demonstrate whether fatty acid synthesis is a necessary component in the sequence of events leading to hormone-stimulated respiration.

The reason why fatty acid cycling occurs in the brown fat tissue but not in other tissues is not clear. The effectiveness of malonyl CoA as an inhibitor of palmitoyl carnitine CoA transferase in brown fat mitochondria has been studied (Saggerson and Carpenter, 1982). Malonyl CoA has a particularly low K_I for the enzyme in brown fat, but the significance of the value obtained ($>0.5\ \mu M$) cannot be evaluated without estimates of the tissue malonyl CoA levels in both the basal and stimulated states. Likewise, it is not known whether there is a protein kinase in brown fat that could phosphorylate and thereby inhibit acetyl CoA carboxylase.

SUMMARY AND CONCLUSIONS

The involvement of the uncoupling protein in the mechanism of action of norepinephrine in brown fat seems well established. Studies of the molecular biology of the protein are underway and promise to provide interesting information about mechanisms of differentiation and adaptation (Ricquier et al., 1983*a*, 1983*b*; Freeman et al., 1983). The reason why the protein is present in such large quantities in the mitochondrial membrane of brown fat is puzzling, especially in view of the fact that activation of a very small proportion of the protein is sufficient to produce functional uncoupling (Nicholls, 1979).

The physiological mechanism for control of the protein remains unresolved. The present reviewer favors the following scheme for the sequence of events following binding of norepinephrine to the β-receptors in the brown fat cell outer membranes. The tissue cyclic AMP levels increase and an endogenous lipase is activated by a phosphorylation mediated by cyclic AMP dependent protein kinase, increasing the level of free fatty acids in the cell enough to activate fatty acyl CoA synthase, whereas the level of free fatty acids in the basal state is high enough to increase proton conductivity mediated by the uncoupling protein. However, in the basal state, ATP levels in the cell cytosol counteract the effect of the fatty acids on the protein. A balance of proton pumping and proton leakage is achieved in the basal steady-state with a $\Delta\Psi_m$ of approximately 120 mvolts.

After hormone treatment, fatty acids and fatty acyl CoA are oxidized by the mitochondria, and citrate levels rise, increasing first in the mitochondria and then in the cytosol. Flux through ATP citrate lyase, acetyl CoA carboxylase, and fatty acyl CoA synthase increases the ATP utilization by these enzymes, thereby lowering cytosolic ATP. These events relieve the inhibition of proton conduction through the uncoupling protein and decrease $\Delta\Psi_m$, allowing a redistribution of cell ATP into the mitochondrial matrix, lowering cytosolic ATP but not total cellular ATP. The increase in respiration follows the decrease in ATP and maximal respiration is achieved at negligible levels of $\Delta\Psi_m$. The reversal of these events is initiated by the decrease in activity of the hormone-sensitive lipase followed by a decrease in the level of free fatty acids and then a decrease in fatty acid substrate cycling. The attendant increase in cytosolic ATP triggers the recoupling of the mitochondria.

Other interpretations are possible, and a scheme in which cytosolic ATP remains relatively high and the protein is regulated not by bound ATP but by the free fatty acid levels in the cytosol is feasible. However, the low $\Delta\Psi_m$ measured in brown fat cells in situ in the basal state does not support the scheme that involves direct regulation by the free fatty acids.

ACKNOWLEDGMENTS

The author wrote this review at the Department of Biochemistry, Oxford University while on sabbatical leave from the Department of Physiology, Pennsylvania State University. Studies reported from the author's laboratory were supported by AM 26309. The author is especially indebted to Dr. John Challis for helpful discussions during the preparation of the manuscript, to Susan Hutson, Charles Koch, and Tad Strzelecki for assistance in the preparation of the manuscript, and to Sue Bennett and Patricia Gering for typing the manuscript.

REFERENCES

Agius, L., and D. H. Williamson, 1980, Lipogenesis in interscapular brown adipose tissue of virgin, pregnant, and lactating rats, *Biochem. J.* **190**:477-480.

Agius, L., D. I. Watts, and M. C. Sugden, 1983, Effect of dichloroacetate on fatty acid synthesis in interscapular brown adipose tissue in the rat *in vivo,* and *in vitro, Horm. Metab. Res.* **15**:482-485.

Agius, L., B. J. Rolls, E. A. Rowe, and D. H. Williamson, 1981, Increased lipogenesis in brown adipose tissue by lactating rats fed a cafeteria diet, *FEBS Lett.* **123:**45-48.

Agius, L., B. J. Rolls, E. A. Rowe, and D. H. Williamson, 1983, High energy diets produce different effects on fatty acid synthesis in brown adipose tissue, white adipose tissue, and liver in the rat, *Biochim. Biophys. Acta* **750:**383-387.

Arslan, P., A. N. Corps, T. R. Hesketh, J. C. Metcalfe, and T. Pozzan, 1984, Cis-unsaturated fatty acids uncouple mitochondria and stimulate glycolysis in intact lymphocytes, *Biochem. J.* **217:**419-425.

Ashwell, M., G. Jennings, D. Richard, D. M. Stirling, and P. Trayburn, 1983, Effect of acclimation temperature on the concentration of the mitochondrial uncoupling protein measured by radioimmunoassay in mouse brown adipose tissue, *FEBS Lett.* **161:**108-112.

Balaban, R.S., L. J. Mandel, S. P. Soltoff, and J. M. Storey, 1980, Coupling of active ion transport and aerobic respiratory rate in isolated renal tubules, *Natl. Acad. Sci. (USA) Proc.* **77:**447-451.

Barnard, T., G. Mory, and M. Nechad, 1980, Biogenic amines and the trophic response of brown adipose tissue, in *Biogenic Amines in Development,* H. Parvez and S. Parvez, eds., Elsevier/North-Holland, Amsterdam, pp. 391-439.

Bieber, L., L. B. Pettersson, and O. Lindberg, 1975, Studies on norepinephrine induced efflux of free fatty acids from hamster brown adipose tissue cells, *Eur. J. Biochem.* **58:**375-381.

Brand, M. D., and S. M. Felber, 1984, Membrane potential of mitochondria in intact lymphocytes during early mitogenic stimulation, *Biochem. J.* **217:**453-459.

Brooks, B. J., J. R. S. Arch, and E. A Newsholme, 1983, Effect of some hormones on the rate of the triacylglycerol fatty acid substrate cycle in adipose tissue of the mouse *in vivo, Bioscience Reports* **3:**263-267.

Bukowiecki, L. N., N. Follea, A. Paradis, and A. Collet, 1980, Stereo-specific stimulation of brown adipocyte respiration by catecholamines via β-adrenoreceptors, *Am. J. Physiol.* **238:**E552-E563.

Bukowiecki, L. J., N. Follea, J. Lupien, and A. Paradis, 1981, Metabolic relationships between lipolysis and respiration in rat brown adipocytes, *J. Biol. Chem.* **256:**12840-12848.

Bulychev, A., R. Kramar, Z. Drahota, and O. Lindberg, 1972, Role of a specific endogeneous fatty acid fraction in the coupling-uncoupling mechanism of oxidative phosphorylation of brown adipose tissue, *Exp. Cell Res.* **72:**169-187.

Caldwell, F. T., Jr., 1962, Metabolic responses to thermal trauma. II. Nutritional studies with rats at two environmental temperatures, *Ann. Surgery* **155:**119-126.

Cannon, B., and G. Vogel, 1977, Mitochondrial ATPase of brown adipose tissue; Purification and comparison with the mitochondrial ATPase from beef heart, *FEBS Lett.* **76:**284-289.

Cannon, B., S. Alexson, and J. Nedergaard, 1982, Peroxisomal oxidation in brown fat, *N. Y. Acad. Sci. Ann.* **386:**40-57.

Cannon, B., J. Nedergaard, and U. Sundin, 1981, Thermogenesis, brown fat, and thermogenin, in *Survival in the Cold, Hibernation and Other Adaptations,* X. Musacchia, and L. Jansky, eds., Elsevier/North Holland, Amsterdam, pp. 99-120.

Cannon, B., D. G. Nicholls, and O. Lindberg, 1973, Purine nucleotides and fatty acids in energy coupling of mitochondria from brown adipose tissue, in *Mechanisms of Bioenergetics,* G. F. Azzone, L. Ernster, S. Papa, E. Quagliariello, and N. Siliprandi, eds., Academic Press, New York, pp. 357-363.

Cannon, B., V. Sundin, and L. Romert, 1977, Palmitoyl CoA: a possible physiological regulation of nucleotide binding to brown adipose tissue mitochondria, *FEBS Lett.* **74:**43-46.

Chakrabarty, K., B. Chaudhure, and H. Jeffay, 1983, Glycerokinase activity in human brown adipose tissue, *J. Lipid Res.* **24:**381-390.

Crass, M. F., III, E. S. McCaskill, and J. C. Shipp, 1969, Effects of pressure development on glucose and palmitate metabolism in perfused heart, *Am. J. Physiol.* **216:**1569-1576.

Davis, E. J., and W. I. A. Davis-van Thienen, 1978, Control of mitochondrial metabolism by the ATP/ADP ratio, *Biochem. Biophys. Res. Commun.* **83:**1260-1266.

Davis, R. J., and B. R. Martin, 1982, The effect of α-adrenergic agonists on the membrane potential of fat cell mitochondria *in situ, Biochem. J.* **206:**619-626.

Desautels, M., and J. Himms-Hagen, 1979, Roles of noradrenalin and protein synthesis in the cold-induced increase in purine by rat brown adipose tissue mitochondria, *Can. J. Biochem.* **57:**968-976.

Desautels, M., G. Zaror-Behrens, and J. Himms-Hagen, 1978, Increased purine nucleotide binding, altered polypeptide composition, and thermogenesis in brown adipose tissue mitochondria of cold-adapted rats, *Can. J. Biochem.* **56:**378-383.

Drahota, Z., E. Hanova, and P. Hahn, 1968, The Effect of ATP and carnitine on the endogenous respiration of mitochondria from brown adipose tissue, *Experientia* **24:**431-432.

Edelman, I. S., 1974, Thyroid thermogenesis, *N. Eng. J. Med.* **209:**1303-1308.

Fain, J. H., and J. A. Garcia-Sainz, 1983, Adrenergic regulation of adipocyte metabolism, *J. Lipid Res.* **24:**945-966.

Fain, J. N., M. D. Jacobs, and Y. C. Clement-Cormier, 1973, Interrelationships of cyclic AMP, liposysis and respiration in brown fat cells, *Am. J. Physiol.* **224:**346-351.

Fain, J. N., N. Reed, and R. Saperstein, 1967, The isolation and metabolism of brown fat cells, *J. Biol. Chem.* **242:**1887-1894.

Ferguson, S. J., and M. C. Sorgato, 1982, Proton electrochemical gradients and energy transduction processes, *Ann. Rev. Biochem.* **51:**185-218.

Fink, S. A., and J. A. Williams, 1976, Adrenergic receptors mediating depolarization in brown adipose tissue, *Am. J. Physiol.* **231:**700-706.

Flaim, K. E., B. A. Horwitz, and J. M. Horowitz, 1977, Coupling of signals to brown fat: α- and β-adrenergic responses in intact rats, *Am. J. Physiol.* **232:**R101-R109.

Fossel, E. T., and A. K. Solomon, 1977, Membrane mediated link between ion transport and metabolism in human red cells, *Biochim. Biophys. Acta* **464:**82-92.

Foster, D. O., and M. L. Frydman, 1978*a,* Comparison of microspheres and ^{86}Rb as tracers of the distribution of cardiac output in rats indicates invalidity of ^{86}Rb-based measurements, *Can. J. Physiol. Pharmacol.* **56:**97-109.

Foster, D. O., and M. L. Frydman, 1978*b,* Non-shivering thermogenesis in the rat II. Measurements of blood flow with microspheres point to brown adipose tissue as the dominant site of calorigenesis induced by nonadrenalin, *Can. J. Physiol. Pharmacol.* **56:**110-122.

Foster, D. O., and M. L. Frydman, 1979, Tissue distribution of cold-induced thermogenesis in conscious warm- or cold-adapted rats; the dominant role of brown adipose tissue, *Can. J. Physiol. Pharmacol.* **57:**257-270.

Freeman, K. B., S. M. Chien, D. Litchfield, and H. V. Patel, 1983, Synthesis *in vitro* of rat brown adipose tissue 32,000 Mr protein, *FEBS Lett.* **158:**325-330.

Garcia-Sainz, J. A., A. K. Hasler, and J. N. Fain, 1980, α-adrenergic activation of phosphoinositol labelling in isolated brown fat cells, *Biochem. Pharmacol.* **29:**3330-3333.

Girardier, L., and G. Schneider-Picard, 1983, α- and β-adrenergic mediation of membrane potential changes and metabolism in rat brown adipose tissue, *J. Physiol.* **335:**629-641.

Girardier, L., J. Seydoux, and T. Clausen, 1968, Membrane potential of brown adipose tissue: a suggested mechanism for the regulation of thermogenesis, *J. Gen. Physiol.* **52:**925-940.

Gnoni, G. V., C. Landriscina, F. M. Ruggiero, and E. Quagliariello, 1983, Effect of hyperthyroidism on lipogenesis in brown adipose tissue of young rats, *Biochim. Biophys. Acta* **751:**271-279.

Goodbody, A. E., and P. Trayhurn, 1982, Studies on the activity of brown adipose tissue in suckling pre-obese (ob/ob) mice, *Biochim. Biophys. Acta* **680:**119-126.

Heaton, G. M., and D. G. Nicholls, 1976, Hamster brown adipose tissue mitochondria: the role of fatty acids in the control of the proton conductance of the inner membrane, *Eur. J. Biochem.* **67:**511-517.

Heaton, G. M., and D. G. Nicholls, 1977, The structural specificity of the nucleotide binding site and the reversible nature of the inhibition of proton conductance induced by bound nucleotides in brown adipose tissue mitochondria, *Biochem. Soc. Trans.* **5:**210-212.

Heaton, G. M., R. J. Wagenvoord, A. Kemp, and D. G. Nicholls, 1978, Brown adipose tissue mitochondria: photoaffinity labelling of the regulatory site for energy dissipation, *Eur. J. Biochem.* **82:**515-521.

Herd, P. A., R. P. Hammond, and M. W. Hamolsky, 1973, Sodium pump activity during norepinephrine stimulated respiration in brown adipocytes, *Am. J. Physiol.* **224:**1300-1304.

Hettinger, D. R., and B. A. Horwitz, 1983, Kinetic characteristics of hamster (Mesocricetus auratics) brown fat Na^+, K^+ ATPase: effects of catecholamines, *Comp. Biochem. Physiol.* **74:**355-360.

Himms-Hagen, J., 1970, Regulation of metabolic processes in brown adipose tissue, *Adv. Enzyme Reg.* **8:**131-151.

Himms-Hagen, J., 1976, Cellular thermogenesis, *Ann. Rev. Physiol.* **38:**315-351.

Himms-Hagen, J., 1979, Obesity may be due to a malfunctioning of brown fat, *Can. Med. Assoc. J.* **17:**1361-1364.

Himms-Hagen, J., and M. Desautels, 1978, A mitochondrial defect in brown adipose tissue of the obese (ob/ob) mouse: reduced binding of purine nucleotides and a failure to respond to cold by an increase in binding, *Biochem. Biophys. Res. Commun.* **83:**628-634.

Himms-Hagen, J., J. Triandafillon, and C. Gilliam, 1981, Brown adipose tissue of cafeterial-fed rats, *Am. J. Physiol.* **241:**E116-E120.

Hittelman, K. J., O. Lindberg and B. Cannon, 1969, Oxidative phosphorylation and compartmentation of fatty acid metabolism in brown fat mitochondria, *Eur. J. Biochem.* **11:**183-192.

Hoek, J. B., D. G. Nicholls, and J. R. Williamson, 1980, Mitochondrial protonmotive force in hepatocytes, *J. Biol. Chem.* **255:**1458-1464.

Hohorst, H. J., and J. Raphael, 1968, Oxidative phosphorylierung durch mitochondrien aus braunem fettgewebe, *Hoppe-Seyler's Z. Physiol. Chem.* **349:**268-270.

Holland, R., L. A. Witters, and D. B. Hardie, 1984, Glucagon inhibits fatty acid synthesis in isolated hepatocytes via phosphorylation of acetyl CoA carboxylase by cyclic AMP dependent protein kinase, *Eur. J. Biochem.* **140:**325-333.

Horowitz, J. M., B. A. Horwitz, and R. E. Smith, 1971, Effect *in vivo* of norepinephrine on the membrane resistance of brown fat cells, *Experientia* **27:**1419-1421.

Horwitz, B. A., 1973, Ouabain-sensitive component of brown fat thermogenesis, *Am. J. Physiol.* **224:**352-355.

Horwitz, B. A., 1979, Cellular events underlying catecholamine-induced thermogenesis: cation transport in brown adipocytes, *Fed. Proc.* **38:**2170-2176.

Horwitz, B. A., and M. Easton, 1975, The effect of adrenergic agonists and cyclic AMP on the (Na + K)-ATPase activity of brown adipose tissue, *Eur. J. Pharmacol.* **34:**241-245.

Horwitz, B. A., J. M. Horowitz, and R. E. Smith, 1969, Norepinephrine-induced depolarization of brown fat cells, *Natl. Acad. Sci. (USA) Proc.* **64:**113-120.

Joseph, S. K., A. P. Thomas, R. J. Williams, R. F. Irvine, and J. R. Williamson, 1984, Myo-inositol-1,4,5-trisphosphate a second messenger for the hormonal mobilization of intracellular Ca^{2+} in liver, *J. Biol. Chem.* **254:**3077-3081.

Klein, R. A., and D. H. Williamson, 1982, Effects of glucose containing peritoneal dialysis solutions on rates of lipogenesis *in vivo* in the liver, brown and white adipose tissue of chronic uraemic rats, *Biochem. J.* **214:**459-464.

Klingenberg, M., 1964, Reversibility of energy transformation in the respiratory chain, *Angewandte Chemie Internationale* **3:**54-61.

Klingenberg, M., and H. Rottenberg, 1977, Relation between the gradient of the ATP/ADP ratio and the membrane potential across the mitochondrial membrane, *Eur. J. Biochem.* **73:**125-130.
Knight, B. L., 1974, Adenosine 3′5′-cyclic phosphate, lipolysis and oxygen consumption in brown adipose tissue from newborn rabbits, *Biochim. Biophys. Acta.* **343:**287-296.
Knight, B. L., and N. B. Myant, 1970, Comparison between the effects of cold exposure *in vivo* and of nonadrenalin *in vitro* on the metabolism of brown fat of newborn rabbits, *Biochem. J.* **119:**103-111.
Knight, B. L., and N. B. Myant, 1971, The effect of noradrenalin on glyceride synthesis and oxidative metabolism *in vitro* on the brown fat of newborn rabbits, *Biochem. J.* **125:**1-8.
Kobayashi, K., and J. R. Neely, 1979, Control of maximum rates of glycolysis in rat cardiac muscle, *Circ. Res.* **44:**166-175.
Kopecky, J., F. Guerrieri, P. J. Zdenek, Z. Drahota, and J. Houstek, 1984, Molecular mechanism of uncoupling in brown adipose tissue mitochondrial: The non-identity of proton and chloride conducting pathways, *FEBS Lett.* **170:**186-190.
Kunz, W., R. Bohnensack, G. Bohme, V. Kuster, G. Letko, and P. Schonfeld, 1981, Relationship between extramitochondrial and intramitochondrial adenine nucleotide systems, *Arch. Biochem. Biophys.* **209:**219-229.
Kunz, W., R. Bohnensack, V. Kuster, G. Letko, G. Bohme, J. Duszynski, and L. Wojtczak, 1983, Control of mitochondrial respiration, *FEBS Lett.* **151:**1-9.
LaNoue, K. F., C. D. Koch, and R. B. Meditz, 1982, Mechanism of action of norepinephrine in hamster brown adipocytes, *J. Biol. Chem.* **257:**13740-13748.
LaNoue, K. F., S. M. Mizani, and M. Klingenberg, 1978, Electrical imbalance of adenine nucleotide transport across the mitochondrial membrane, *J. Biol. Chem.* **253:**191-198.
LaNoue, K. F., and T. Strzelecki, 1983, Hormonal effects on mitochondrial membrane potentials, *Fed. Proc.* **42:**2265.
LaNoue, K. F., T. Strzelecki, and F. Finch, 1984, The effect of glucagon on hepatic respiratory capacity, *J. Biol. Chem.* **259:**4116-4121.
LaNoue, K. F., D. Strzelecki, and C. D. Koch, 1984, Na^+, K^+-ATPase activity and ion fluxes in intact brown adipocytes, *Biophys. J.* **45:**159a.
LaNoue, K. F., T. Strzelecki, D. Strzelecka, and C. D. Koch, 1986, Regulation of the uncoupler protein in brown adipose tissue, *J. Biol. Chem.* **261:**298-305.
LaNoue, K. F., C. D. Koch, D. Strzelecka, and T. Kobylski, 1986, Regulation of Na^+ transport in brown adipose tissue, *Biochem. J.* (in press).
Lawford, H. G., and P. B. Garland, 1972, Proton translocation coupled to quinone reduction by reduced nicotinamide-adenine dinucleotide in rat liver and ox heart mitochondria, *Biochem. J.* **130:**1029-1044.
Lean, M. E., and W. P. James, 1983, Uncoupling protein in human brown adipose tissue mitochondria. Isolation and detection by specific antisera, *FEBS Lett.* **163:**235-240.
Lehninger, A. L., 1975, The molecular basis of cell structure and function in *Biochemistry,* 2nd ed., Worth Publishers, New York, pp.387-413.
Lin, C. S., H. Hackenberg, and M. Klingenberg, 1980, The uncoupling protein from brown adipose tissue mitochondria is a dimer: a hydrodynamic study, *FEBS Lett.* **113:**304-306.
Lin, C. S., and M. Klingenberg, 1980, Isolation of the uncoupling protein from brown adipose tissue mitochondria, *Biochemistry* **21:**2950-2956.
Lin, C. S., and M. Klingenberg, 1982, Characteristic of the isolated purine nucleotide binding protein from brown fat mitochondria, *Biochemistry* **21:**2950-2956.
Lindberg, O. (ed.), 1970, *Brown Adipose Tissue,* Elsevier, New York.
Lindberg, O., J. Nedergaard, and B. Cannon, 1981, Thermogenic mitochondria, in *Mitochondria and Microsomes,* C. P. Lee, G. Schatz, and G. Dallner, eds., Addison-Wesley, Menlo Park, Calif., pp. 93-119.

Lindberg, O., J. DePierre, E. Rylander, and B. A. Afzalins, 1967, Studies of the mitochondrial energy transfer system of brown adipose tissue, *J. Cell Biol.* **34:**293-310.

Locke, R. M., and D. G. Nicholls, 1981, A re-evaluation of the role of fatty acids in the physiological regulation of the proton conductance of brown adipose tissue mitochondria, *FEBS Lett.* **135:**249-252.

Locke, R. M., E. Rial, and D. G. Nicholls, 1982, The acute regulation of mitochondrial proton conductance in cell and mitochondria from the brown fat of cold- and warm-adapted guinea pigs, *Eur. J. Biochem.* **129:**381-387.

Locke, R. M., E. Rial, I. D. Scott, and D. G. Nicholls, 1982, Fatty acids as acute regulators of the proton conductance of hamster brown fat mitochondria, *Eur. J. Biochem.* **129:**373-380.

Maccechini, M. L., Y. Rudin, G. Blobel, and G. Schatz, 1979, Import of proteins into mitochondrial: Precursor forms of the extramitochondrially made F_i-ATPase subunits in yeast, *Natl. Acad. Sci. (USA) Proc.* **76:**343-347.

McCormack, J. G., 1982, The regulation of fatty acid synthesis in brown adipose tissue by insulin, *Prog. Lipid Res.* **21:**195-223.

McCormack, J. G., and R. M. Denton, 1977, Evidence that fatty acid synthesis in the interscapular brown adipose tissue of cold adapted rats is increased *in vivo* by insulin by mechanisms involving parallel activation of pyruvate dehydrogenase and acetyl CoA carboxylase, *Biochem. J.* **166:**627-630.

McGarry, J. D., and D. W. Foster, 1979, In support of the roles of malomyl CoA and carnitine acyltransferase I in the regulation of hepatic fatty acid oxidation and ketogenesis, *J. Biol. Chem.* **254:**8163-8168.

McGarry, J. D., and D. W. Foster, 1980, Regulation of hepatic fatty acid oxidation and ketone body production, *Annu. Rev. Biochem.* **49:**395-420.

Marchall, C. E., D. I. Watts, and M. C. Sugden, 1983, Effect of hydrazine in liver and brown adipose tissue lipogenesis in 24-hour starved rats, *J. Pharm. Pharmacol.* **35:**460-461.

Masoro, E. J., 1963, Role of lipogenesis in nonshivering thermogenesis, *Fed. Proc.* **22:**868-873.

Mitchell, P., 1976, Vectorial chemistry and the molecular mechanisms of chemiosmotic coupling, *Biochem. Soc. Trans.* **4:**399-430.

Mohell, N., J. Nedergaard, and B. Cannon, 1983, Quantitative differentiation of α- and β-adrenergic respiratory responses in isolated hamster brown fat cells: evidence for the presence of an α-adrenergic component, *Eur. J. Pharmacol.* **93:**183-193.

Mohell, N., J. Svatengren, and B. Cannon, 1983, Identification of [^{3}H]prazosin binding sites in crude membranes and isolated cells of brown adipose tissue as α-1 adrenergic receptors, *Eur. J. Pharmacol.* **92:**15-25.

Mohell, N., M. Wallace, and J. N. Fain, 1984, Alpha-adrenergic stimulation of phosphatidylinositol turnover and respiration of brown fat cells, *Mol. Pharmacol.* **25:**64-69.

Nedergaard, J., 1981, Effects of cations on brown adipose tissue in relation to possible metabolic consequences of membrane depolarization, *Eur. J. Biochem.* **114:**159-167.

Nedergaard, J., 1983, Relationship between extramitochondrial Ca^{++} concentration, respiratory rate, and membrane potential in mitochondria of brown adipose tissue of the rat, *Eur. J. Biochem.* **133:**185-191.

Nedergaard, J., and O. Lindberg, 1979, Norepinephrine-stimulated fatty acid release and oxygen consumption in isolated brown fat cells. Influence of buffers, albumin, insulin and mitochondrial inhibitors, *Eur. J. Biochem.* **95:**139-145.

Nedergaard, J., and O. Lindberg, 1982, The brown fat cell, *Internat. Rev. Cytol.* **74:**187-286.

Newsholme, E. A., 1980, A possible metabolic basis for the control of body weight, *N. Eng. J. Med.* **302:**400-405.

Newsholme, E. A., R. A. J. Challiss, and B. Crabtree, 1984, Substrate cycles: their role in improving sensitivity in metabolic control, *Trends Biochem. Sci.* **9:**277-280.

Newsholme, E. A., and B. Crabtree, 1976, Substrate cycles in metabolic regulation and in heat generation. *Biochem. Soc. Symp.* **41:**61-109.

Nicholls, D. G., 1974*a,* Hamster brown adipose tissue mitochondria: the content of respiration and the proton electrochemical potential gradient by possible physiological effectors of the proton conductance of the inner membrane, *Eur. J. Biochem.* **49:**573-583.

Nicholls, D. G., 1974*b,* Hamster brown adipose tissue mitochondria: The chloride permeability of the inner membrane under respiring conditions, the influence of purine nucleotides, *Eur. J. Biochem.* **49:**585-593.

Nicholls, D. G., 1976*a,* Hamster brown adipose tissue mitochondria: purine nucleotide control of the ionic conductance of the inner membrane, the nature of the nucleotide binding site, *Eur. J. Biochem.* **62:**223-228.

Nicholls, D. G., 1976*b,* Brown fat mitochondria, *Trends Biochem. Sci.* **1:**128-130.

Nicholls, D. G., 1977, The effective proton conductances of the inner membrane of mitochondrial from brown adipose tissue depending on proton electrochemical gradient, *Eur. J. Biochem.* **77:**349-356.

Nicholls, D. G., 1979, Brown adipose tissue. *Biochim. Biophys. Acta* **549:**1-22.

Nicholls, D. G., 1982, *Bioenergetics: An Introduction to the chemiosmotic theory,* Academic Press, London.

Nicholls, D. G., and V. S. M. Bernson, 1977, Inter-relationships between $\Delta\mu H^+$, ΔG_p, and respiration during substrate-level and oxidative phosphorylation by mitochondria from brown adipose tissue of cold adapted guinea pigs, *Eur. J. Biochem.* **75:**601-612.

Nicholls, D. G., H. J. Grav, and O. Lindberg, 1972, Mitochondria from brown adipose tissue: regulation of respiration *in vitro* by variations in volume of the matrix compartment, *Eur. J. Biochem.* **31:**526-533.

Nicholls, D. G., and R. M. Locke, 1984, Thermogenic mechanisms in brown fat, *Physiol. Rev.* **64:**1-62.

Oram, J. F., S. L. Bennetch, and J. R. Neely, 1973, Regulation of fatty acid utilization in isolated perfused rat hearts, *J. Biol. Chem.* **248:**5299-5309.

Path, E. A., D. M. Salmon, and D. A. Hems, 1979, Effect of acute change in ambient temperature on fatty acid synthesis in the mouse, *FEBS Lett.* **108:**33-36.

Pettersson, B., 1977, Carbon dioxide-mediated control of fatty acid metabolism in isolated hamster brown fat cells during norepinephrine stimulation, *Eur. J. Biochem.* **72:**235-240.

Pettersson, B., and I. Vallin, 1976, Norepinephrine-induced shift in levels of adenosine 3, '5-monophosphate and ATP parallel to increased respiratory rate and lipolysis in isolated hamster brown fat cells, *Eur. J. Biochem.* **62:**383-390.

Pressman, B. C., and H. A. Lardy, 1956, Effects of surface active agents on latent ATPase of mitochondria, *Biochim. Biophys. Acta* **21:**458-466.

Prusiner, S. B., 1970, Spectroscopic evidence for the control of respiration prior to phosphorylation in hamster brown fat cells, *J. Biol. Chem.* **245:**382-389.

Prusiner, S. B., B. Cannon, and O. Lindberg, 1968, Oxidative metabolism in cells isolated from brown adipose tissue. 1. Catecholamine and fatty acid stimulation of respiration, *Eur. J. Biochem.* **6:**15-22.

Prusiner, S. J., J. R. Williamson, B. Chance, and B. M. Paddle, 1968, Pyridine nucleotide changes during thermogenesis in brown fat tissue *in vivo, Arch. Biochem. Biophys.* **123:**368-377.

Raphael, J., and D. G. Nicholls, 1984, Mitochondrial membrane potential monitored *in situ* within isolated guinea pig brown adipocytes by a styrylpyridinium fluorescent indicator, *FEBS Lett.* **170:**181-185.

Raphael, J., and E. Wrabitz, 1976, Brown adipose tissue mitochondria: recoupling caused by substrate level phosphorylation and extramitochondrial adenosine phosphates, *Eur. J. Biochem.* **61:**551-561.

Raphael, J., D. Klaas, and H. J. Hohorst, 1968, Mitochondria from brown fat: enzymes and

respiratory chain phosphorylation during the pre- and postnatal development of the intrascapular fat body of the guinea pig. *Hoppe-Seyler's Z. Physiol Chem.* **349:**1711-1724.
Raphael, J., H. J. Ludolph, and H. J. Hohorst, 1969, Mitochondria from brown adipose tissue: uncoupling of oxidative phosphorylation by long chain fatty acids and recoupling by guanosine triphosphate, *Hoppe-Seyler's Z. Physiol. Chem.* **350:**1121-1131.
Resh, M. D., R. A. Nemenoff, and G. Guidotti, 1980, Insulin stimulation of (Na^+, K^+)-adenosine triphosphatase-dependent Rb^+ uptake in rat adipocyte, *J. Biol. Chem.* **255:**10938-10945.
Rial, E., and D. G. Nicholls, 1984, The mitochondrial uncoupling protein from guinea-pig brown adipose tissue, *Biochem J.* **222:**685-693.
Ricquier, D., and J.-C. Kader, 1976, Mitochondrial protein alteration in active brown fat: a sodium dodecyl sulphate polyacrylamide electrophoretic study, *Biochem. Biophys. Res. Commun.* **73:**577-583.
Ricquier, D., C. S. Lin, and M. Klingenberg, 1982, Isolation of the GDP binding protein from brown adipose tissue mitochondria of several animals and amino acid composition study in rat, *Biochem. Biophys. Res. Commun.* **106:**582-589.
Ricquier, D., G. Mory, and P. Hemon, 1979, Changes induced by cold adaptation in the brown adipose tissue from several species of rodents with special reference to the mitochondrial components, *Can. J. Biochem.* **57:**1262-1266.
Ricquier, D., M. Nechad, and G. Mory, 1982, Ultrastructure and biochemical characterization of human brown adipose tissue in Pheochromocytoma, *J. Clin. Endocrinol. Metab.* **54:**803-807.
Ricquier, D., C. Gervais, J.-C. Kader, and P. Hemon, 1980, Partial perufication by GDP-agarose affinity chromatography of the 32,000 M polypeptide from mitochondria of brown adipose tissue, *FEBS Lett.* **101:**35-38.
Ricquier, D., J. Thibault, F. Bouilland, and Y. Kuster, 1983*a*, Molecular approach to thermogenesis in brown adipose tissue: cell free translation of a RNA and characterization of the mitochondria uncoupling protein, *J. Biol. Chem.* **258:**6675-6677.
Ricquier, D., J. P. Bartlett, J. M. Garel, M. Combes-George, and M. P. Dubois, 1983*b*, An immunological study of the uncoupling protein of brown adipose tissue mitochondria, *Biochem. J.* **210:**859-866.
Rothwell, N. J., and M. J. Stock, 1979, A role for brown adipose tissue in diet-induced thermogenesis, *Nature* **281:**31-53.
Rothwell, N. J., and M. J. Stock, 1980, Similarities between cold- and diet-induced thermogenesis in the rat, *Can. J. Physiol. Pharmacol.* **58:**842-848.
Rothwell, N. J., and M. J. Stock, 1981, Regulation of energy balance, *Annu. Rev. Nutr.* **1:**235-256.
Rothwell, N. J., and M. J. Stock, 1982, Neural regulation of thermogenesis, *Trends Neurosci.* **5:**124-127.
Rothwell, N. J., M. J. Stock, and D. Stribling, 1982, Diet-induced thermogenesis, *Pharmacol. Ther.* **17:**251-268.
Rothwell, N. J., M. J. Stock, and M. G. Wyllie, 1981, Na^+,K^+ ATPase activity and noradrenalin turnover in brown adipose tissue of rats exhibiting diet-induced thermogenesis, *Biochem. Pharmacol.* **30:**1709-1712.
Rottenberg, H., 1983, Uncoupling of oxidative phosphorylation in rat liver mitochondria by general anesthetics, *Natl. Acad. Sci. (USA) Proc.* **80:**3313-3317.
Rottenberg, H., 1984, Membrane potential and surface potential in mitochondria: Uptake and binding of lipophilic cations, *J. Memb. Biol.* **81:**127-138.
Ruzicka, F. I., and F. L. Crane, 1971, Quinone interaction with the respiratory chain-linked NADH dihydrogenease of beef heart mitochondria. II. Duroquinone reductase activity. *Biochim. Biophys. Acta* **226:**221-233.
Saggerson, E. D., and C. A. Carpenter, 1982, Sensitivity of brown-adipose tissue carnitine palmitoyl transferase in inhibition by malonyl CoA, *Biochem. J.* **204:**373-375.
Scott, I. D., and D. G. Nicholls, 1980, Energy translocation in intact synaptosomes: influence

of plasma membrane depolorization on the respiration and membrane potential of internal mitochondria determine *in situ, Biochem. J.* **182:**21-33.

Senior, A. E., and J. G. Wise, 1983, The proton-ATP-ase of bacteria and mitochondria, *J. Membrane Biol.* **73:**105-124.

Seydoux, J., F. Assimacopoulos-Jeannet, R. Jeanrenaud, and L. Girardier, 1982, Alterations of brown adipose tissue in genetically obese (Ob/Ob) mice. I. Demonstration of loss of metabolic response to nerve stimulation and catecholamines and its partial recovery after fasting or cold adaptation, *Endocrinology* **110:**432-438.

Seydoux, J., A. Chinet, G. Schneider-Picard, S. Bas, E. Imesch, F. Assimacopoulos-Jeannet, J. P. Giacobino, and L. Girardier, 1983, Brown adipose tissue metabolism in streptozotocin-diabetic rats, *Endocrinology* **113:**604-610.

Skala, J. P., and B. L. Knight, 1979, Cyclic GMP and cyclic GMP-dependent protein kinase in brown adipose tissue of developing rats, *Biochim. Biophys. Acta* **582:**122-131.

Smith, R. E., and B. A. Horwitz, 1969, Brown fat and thermogenesis, *Physiol. Rev.* **49:**330-425.

Smith, R. E., J. C. Roberts, and K. J. Hittelman, 1966, Non-phosphorylating respiration of mitochondria from brown adipose tissue of rats, *Science* **154:**653-654.

Streb, H., R. F. Irvine, M. J. Berridge, and I. Shulz, 1983, Release of Ca^{2+} from a nonmitochondrial intracellular store in pancreatic acinar cells by inositol-1,4,5-triphosphate, *Nature* **306:**67-69.

Strieleman, P. J., K. L. Schalinske, and E. Shrago, 1985, Partial purification and functional reconstitution of GDP sensitive brown adipose tissue mitochondria uncoupling protein using octyl glucoside, *Biochem. Biophys. Res. Commun.* **127:**509-516.

Strunecka, A., L. Olivierusova, V. Kubista, and Z. Drahota, 1981, Effect of cold stress and norepinephrine on the turnover of phospholipids in brown adipose tissue of the golden hamster, *Physiol. Bohemoslov* **30:**307-313.

Sugden, M. C., D. I. Watts, C. E. Marshall, and J. G. McCormack 1982, Brown adipose tissue lipogenesis in starvation: effects of insulin and (−) hydroxycitrate, *Biosci. Rep.* **2:**289-297.

Sugden, M. C., S. E. Steare, D. I. Watts, and T. N. Palmer, 1983, Interactions between insulin and thyroid hormone on the control of lipogenesis, *Biochem. J.* **210:**937-944.

Sundin, U., and J. N. Fain, 1983, α-adrenergic inhibition of lipolysis and respiration in rat brown adipocytes, *Biochem. Pharmacol.* **32:**3117-3120.

Svoboda, P. J., M. Svartengren, J. Houstek Snochowski, and B. Cannon, 1979, High number of high affinity binding sites for (-)-dihydroalprenolol on isolated hamster brown fat cells, *Eur. J. Biochem.* **102:**203-210.

Tokuyama, K., and H. Kuda, 1983, Fatty acid synthesis in adipose tissues of physically brained rats *in vivo, Am. J. Physiol.* **245:**E8-E13.

Trayhurn, P., 1979, Fatty acid synthesis *in vivo* in brown adipose tissue, liver and white adipose tissue of the cold adapted rat, *FEBS Lett.* **104:**13-16.

Trayhurn, P., 1980, Fatty acid synthesis in brown adipose tissue in relation to whole body synthesis in the cold-adapted golden hamster, *Biochim. Biophys. Acta* **620:**10-17.

Trayhurn, P., 1981, Fatty acid synthesis in mouse brown adipose tissue, *Biochim. Biophys. Acta* **664:**549-560.

Trayhurn, P., and D. R. Fraser, 1983, The influence of GDP on Ca^{2+} uptake by mitochondria of brown adipose tissue from lean and genetically obese (Ob/Ob) mice, *Biochem. J.* **214:**171-175.

Van den Brandt, P. A., and P. Trayhurn, 1981, Suppression of fatty acids synthesis in brown adipose tissue of mice fed diets rich in long chain fatty acids, *Biochim. Biophys. Acta* **665:**602-607.

Watts, J. A., C. D. Koch, and K. F. LaNoue, 1980. Effects of Ca^{2+} antagonism on energy metabolism: Ca^{2+} and heart function after ischemia, *Am. J. Physiol.* **238:**N909-N916.

Wikstrom, J., K. Krab, and M. Saraste, 1981, Proton translocating cytochrome complexes, *Ann. Rev. Biochem.* **50:**623-655.

Williams, J. A., and E. K. Matthews, 1974*a,* Effect of ions and metabolic inhibitors on membrane potential of brown adipose tissue, *Am. J. Physiol.* **227**:981-986.
Williams, J. A., and E. K. Matthews, 1974*b,* Membrane depolarization, cyclic AMP, and glycerol release by brown adipose tissue, *Am. J. Physiol.* **227**:987-992.
Williamson, J. R., 1970, Control of energy metabolism in hamster brown adipose tissue, *J. Biol. Chem.* **245**:2043-2050.
Witter, L. A., J. P. Tipper, and G. W. Bacon, 1983, Stimulation of site-specific phosphorylation of acetyl CoA carboxylase by insulin and epinephrine, *J. Biol. Chem.* **258**:5643-5648.
Wright, J., and L. Agius, 1983, Fatty acid synthesis and ketone body utilization by brown adipose tissue of rat. Response to cold or nutritional state, *Biochim. Biophys. Acta* **753**:244-248.
Zutzinski, C., M. Schleyer, and W. Neupert, 1983, Transfer of proteins into mitochondria: precursor to the ADP/ATP carrier binds to receptor sites on isolated mitochondria, *J. Biol. Chem.* **258**:4071-4074.

2

Mitochondrial Proteolysis

Michel Desautels

University of Saskatchewan

Mitochondria are highly dynamic structures that probably participate in more metabolic functions than any other organelle in the cell. Their number, size, shape, and the abundance of their cristae as well as their protein and enzyme composition differ widely depending upon the specific function and the degree of physiological activity of a tissue or organ. In certain organs they are by far the most prominent organelles. For example, in brown adipose tissue, a small thermogenic organ whose main function is to oxidize fatty acids in order to produce heat, approximately 10,000 mitochondria are within one cell and contain at least one third of the total amount of protein in that cell (Nedergaard and Lindberg, 1982). In cardiac muscle, which derives its energy almost exclusively from aerobic metabolism, as well as in liver, mitochondrial proteins constitute 10 to 15% of the total protein content of the cell (Ernster and Schatz, 1981).

The protein constituents of mitochondria are in a state of dynamic equilibrium, constantly being destroyed and resynthesized during the lifetime of a cell. In heart and brown adipose tissue, where mitochondria are most abundant, this constant replacement of mitochondrial proteins may constitute an important fraction of the basal turnover of proteins. It is evident that protein degradation, like protein synthesis, is essential for the maintenance of a stable and functional mitochondrial mass. Mitochondria and mitochondrial components unable to carry out their normal functions of energy transformations appear to be rapidly eliminated (Luzikov, Makhlis, and Galkin, 1976; Luzikov, 1981). For example, during reticulocyte maturation, mitochondria undergo a process of aging and disappear from the cell (Gasko and Danon, 1972). In vivo hypoxia induces loss of heart mitochondrial components such as cytochrome c and aa_3 (Aschenbrenner et al., 1971; Mela, Goodwin, and Miller, 1976). Similar results are observed in yeasts upon de-aeration of their growth medium (Luzikov, et al., 1971). Destruction of mitochondrial structure and function is known to accompany cellular damage. Complete disappearance of inner membrane structures occurs in liver, muscle, and kidney mitochondria following endotoxemia, hemorrhage, and ischemia (Mela et al., 1973).

In addition, there are a number of physiological situations in which alterations in the rate of mitochondrial protein degradation contribute to markedly changing the cellular content of mitochondrial proteins. For example, acute changes in environmental temperature, food intake, or diet composition are known to alter the rate constant of degradation of liver (Chee and Swick, 1976; Stocco, Cascarano, and Wilson, 1976; Nicoletti, Guerri, and Grisolia, 1977) and brown adipose tissue mitochondria (Bukowiecki and Himms-Hagen, 1971; Desautels and Yuen, 1984; Desautels and Himms-Hagen, 1980; Himms-Hagen, Dittmar, and Zaror-Behrens, 1980) as well as the turnover rate of some of their enzymes (Chee and Swick, 1976; Nicoletti, Guerri, and Grisolia, 1977; Swick, Rexroth, and Stange, 1968). Furthermore, mitochondrial degradation, like mitochondrial biogenesis, appears to be an important process also regulated by various hormonal agents such as corticotrophin in the adrenal cortex (Neri et al., 1978; Nussdorfer, Neri, and Mazzochi, 1978; Nussdorfer et al., 1975) or thyroid hormone in the liver (Nicoletti, Guerri, and Grisolia, 1977; Rajwade et al., 1975). Although there have been considerable advances in our knowledge of the biogenesis of mitochondria and of how mitochondrial proteins coded either by the nuclear or mitochondrial genome are synthesized, processed, and coordinately assembled to form large enzyme complexes, there has been little attention paid to the mechanisms by which whole mitochondria or individual proteins can be degraded in mammalian cells.

The turnover of mitochondrial proteins could occur via one of two processes. Either entire mitochondria could be degraded and replaced (alterations of this balance would lead to changes in the total number of mitochondria in the cell), or proteins could be inserted and removed from an existing population of mitochondria while maintaining the mitochondrial structures intact. Such balance between synthesis and degradation of specific mitochondrial proteins may be crucial for allowing changes in regulatory enzyme levels in response to hormonal stimuli and nutritional and environmental influences. These two processes are not mutually exclusive but may complement each other in causing rapid changes in mitochondrial content and composition. However, under basal conditions, in which mitochondrial proteins are constantly renewed, controversy exists as to whether mitochondria are replaced as entire units or by independent proteolytic systems acting selectively on different mitochondrial constituents. If mitochondria are turning over as whole units, all the proteins that constitute the various compartments (inner and outer membranes, intermembrane, and matrix spaces) should have equivalent half lives. By contrast, heterogeneity in the turnover rates of mitochondrial proteins would suggest the presence of distinct proteolytic systems.

HETEROGENEITY OF TURNOVER RATES

Mitochondrial protein turnover is usually measured by the incorporation and subsequent loss of labeled precursors from the mitochondrial proteins. As with all studies on protein turnover, progress has been largely impaired by the problems inherent in the method of measurement, of which the most acute problem is certainly the re-utilization of labeled precursors derived from protein breakdown. It is beyond the scope of this review to comment on the various questions pertinent to the evaluation of rates of protein turnover. For review, the readers are referred to excellent papers by Goldberg and Dice (1974) and Zak, Martin, and Blough (1979).

Early studies indicated that mitochondria turn over as complete units. In rat liver, similar half lives of 6 to 10 days were observed for DNA, lipids, and a variety of protein fractions. However, the problem of re-utilization of labeled precursors greatly complicated the interpretation of these results (Burgess, Walker, and Mayer, 1978; Glass and Doyle, 1972). In addition, the possibility that similar mean turnover rates for DNA, protein, and lipids were fortuitous could not be completely discounted.

More recent studies using minimally re-utilizable labeled precursors have challenged this earlier view (Aschenbrenner et al., 1970; Gross, Getz, and Rabinowitz, 1969; Swick, Rexroth, and Stange, 1968; Burgess, Walker, and Mayer, 1978; Glass and Doyle, 1972; Walker, Burgess, and Mayer, 1978; Ip, Chee, and Swick, 1974). In adult or weaning rats, a mean half life for liver mitochondrial proteins of 3 to 5 days has been reported (Mayer, 1979; Lipsky and Pedersen, 1981*a,* 1981*b;* Russell, Burgess, and Mayer, 1980). This value, which is much less than the 6 to 10 days originally reported, is identical to that observed in hepatocyte monolayer cultures, to which large amounts of unlabeled precursors can be added to minimize or even abolish the re-utilization of labeled precursors (Evans and Mayer, 1983). Both in tissue cultures or in vivo, the mean half lives of proteins in different mitochondrial compartments are not identical to that of the bulk of mitochondrial proteins. There is general agreement that outer membrane proteins turn over faster than proteins embedded in the inner membrane or present in the matrix

(Mayer, 1979; Lipsky and Pedersen, 1981*a,* 1981*b;* Druyan, Debernard, and Rabinowitz, 1969; Gear, 1970). In addition, heterogeneity of turnover rates of outer membrane proteins has been observed. Two mean populations of degradative rates (1.9 days and >4.0 days) have been reported (Russell, Burgess, and Mayer, 1982), and certain outer membrane enzymes turn over at widely distinct rates. For example, ornithine decarboxylase has a half life of less than 20 minutes (Dice and Goldberg, 1975) while monoamine oxidase has a half life of 55 to 60 hours (Evans and Mayer, 1983, 1984) and NADH-cytochrome b_5 reductase, 128 hours (Okada and Omura, 1978).

The rate of degradation of composite inner membrane proteins is no different from that of the matrix proteins and it has been suggested that the inner compartment (consisting of inner membrane and matrix) may be degraded as a unit independently of the outer compartments (Gross, Getz, and Rabinowitz, 1969; Lipsky and Pedersen, 1981*a,* 1981*b;* Russell, Burgess, and Mayer, 1980). However, heterogeneity of degradative rates exists for individual proteins in all mitochondrial compartments. There is general agreement that mitochondrial matrix polypeptides can be inserted and removed at heterogeneous rates. For example, in hepatoma monolayer cultures, 33 mitochondrial matrix polypeptides resolved by two-dimensional gel electrophoresis show half lives ranging from 17 to over 100 hours (Hare and Hodges, 1982*a*). Similar heterogeneity of turnover rates is observed for polypeptides present in liver mitochondrial matrix (Mayer, 1979; Walker, Burgess, and Mayer, 1978; Russell, Burgess, and Mayer, 1982). Striking differences also exist in the turnover rates of well characterized matrix enzymes. δ-amino levulinate synthase has a very short half life of only 70 minutes (Dice and Goldberg, 1975). Alanine dehydrogenase, glutamate dehydrogenase, and ornithine aminotransferase have half lives of 18, 24, and 48 hours respectively while carbamoyl phosphate synthase, aspartate amino transferase, fumarase, and malate dehydrogenase have much slower turnover rates with half lives of 3 to 10 days (Mayer, 1979; Nicoletti, Guerri, and Grisolia, 1977; Swick, Rexroth, and Stange, 1968; Ip, Chee, and Swick, 1974; Kobayashi, Kamimura, and Tuboi, 1982; Shaffer and Felder, 1983). These results strongly suggest that the turnover of mitochondrial matrix polypeptides is regulated by a proteolytic system that is capable of discriminating between different matrix polypeptides.

By contrast, the extent of heterogeneity of turnover rates of proteins embedded in the inner mitochondrial membrane is still a subject of debate. Generally it is true that the turnover of integral membrane proteins is significantly slower than that of soluble proteins, and there appears to be less heterogeneity in their rates of degradation (Walker, Burgess, and Mayer, 1978). Early studies showed similar turnover rates for cytochrome b, c, and cytochrome oxidase (Aschenbrenner et al., 1970; Druyan, Debernard, and

Rabinowitz, 1969; Ip, Chee, and Swick, 1974). However, Russell, Burgess, and Mayer (1982) observed heterogeneity in the degradation of inner membrane polypeptides in rat liver mitochondria and noted that these degradative rates could be grouped into two or three mean populations, suggesting limited heterogeneity and possibly coordinate turnover of some of the mitochondrial membrane proteins. In hepatoma monolayer cultures, Hare and Hodges (1982*b*) observed half lives ranging from 20 to over 100 hours for polypeptides associated with the inner membrane, leading them to conclude that mammalian mitochondria may be susceptible to mechanisms of protein breakdown that result in mosaic turnover of the inner membrane.

Mitochondrial translation products, which are for the most part associated with inner membrane enzyme complexes, are degraded heterogeneously with half lives ranging from 30 to over 100 hours (Hare and Hodges, 1982*b*). These different turnover rates are probably a reflection of the degradative rates of the enzyme complexes to which the mitochondrially synthesized polypeptides are associated. In the inner membrane, cytochrome oxidase is very stable and has a turnover rate similar to that of most inner membrane proteins, with a half life greater than five days in liver (Russell, Burgess, and Mayer, 1980; Ip, Chee, and Swick, 1974) and in hepatomas (Hare and Hodges, 1982*b*). Cytochrome oxidase is made up of seven subunits, three of which are synthesized within the mitochondria (Ernster and Schatz, 1981; Schatz, 1981). All the subunits of cytochrome oxidase appear to turn over at a similar rate (Russell, Burgess, and Mayer, 1980; Hare and Hodges, 1982*b*). The same is true of F_1-ATPase, which has a very slow turnover rate, and none of its five subunits are degraded significantly during a 48-hour chase period (Hare and Hodges, 1982*b*). By contrast, the cytochrome b-c_1 complex turns over at a rate faster than cytochrome oxidase and F_1-ATPase (Hare and Hodges, 1982*b*). The turnover rate of pyruvate dehydrogenase is also significantly faster than that of cytochrome oxidase or F_1-ATPase, but all of its subunits turn over at the same rate (Russell, Burgess, and Mayer, 1980). Thus, although there are some differences in the rates of degradation of membrane enzyme complexes, the protein subunits that constitute an individual enzyme complex appear to turn over coordinately. One exception has been found—two of the subunits of the cytochrome b-c_1 complex turn over more slowly ($t_{1/2} = 40$ hours) than its remaining four subunits ($t_{1/2} > 94$ hours) (Hare and Hodges, 1982*b*).

In conclusion, there is no doubt that soluble mitochondrial enzymes turn over at distinct rates. There is also heterogeneity in the degradation of membrane enzymes but more studies are required to evaluate clearly the extent of this heterogeneity. The apparent coordinate turnover of the protein subunits that constitute an inner membrane enzyme complex raises interesting questions. Although it appears logical that such proteins should turn over at similar rates, the mechanism of the degradation of proteins that are probably embedded to different extents in the membrane or that may face either the outer or the inner face of the membrane is rather puzzling.

Whether a mitochondrial protein is more or less susceptible to degradation may depend upon several factors. Obviously, some structural characteristics of the proteins themselves are important in determining their rates of degradation. In the cell cytosol, large proteins or polypeptides with acidic isoelectric points are more susceptible to proteolytic cleavages than small or basic proteins are (Rajwade et al., 1975; Dice and Goldberg, 1975). By contrast, in mitochondria there is no evidence for such a relationship between subunit size or isoelectric point and rates of protein degradation in any of the compartments (Hare and Hodges, 1982*a,* 1982*b;* Walker, Burgess, and Mayer, 1978; Russell, Burgess, and Mayer, 1982). However, it was shown that NADH-cytochrome b_5 reductase, which is present in both microsomal and outer mitochondrial membranes, has an identical rate of degradation, although the turnover time of the outer mitochondrial membrane is significantly longer than that of microsomes (Okada and Omura, 1978). As the composition and functions of the mitochondrial outer membrane differ from those of microsomes, the identical turnover rate of the reductase associated with these two membranes suggests that its degradation depends upon characteristics of the enzyme itself rather than of the membranes. These observations differ from those obtained earlier by Druyan, Debernard, and Rabinowitz (1969) in which cytochrome b_5 had a longer life span in the outer mitochondrial membrane than in microsomes. As cytochrome b_5 may be associated with several flavoproteins, its degradation rate is possibly linked to the turnover of these enzymes.

As in the cell cytosol, it is likely that the degradation of mitochondrial proteins may be altered by binding various cofactors or ligands or by association with other proteins or membrane structures in the organelle. For example, proteolytic inactivation of ornithine transcarbamylase, a matrix enzyme, is markedly reduced in the presence of its substrate, carbamoyl phosphate (Navarro and Grisolia, 1984). Similarly, pyridoxal phosphate significantly reduces the susceptibility of δ-amino levulinate synthase, a pyridoxal phosphate-containing enzyme, to proteolytic inactivation. In fact, the great sensitivity of the apo-enzyme to proteases (Aoki et al., 1975) may explain its low levels in erythroblasts of patients with pyridoxine-responsive anemia or in tissues of pyridoxine-deficient animals (Aoki et al., 1979). Another example is ornithine aminotransferase, which appears to be stabilized by the presence of L-tryptophan (Chee and Swick, 1976). By contrast, NADP potentiates the inactivation of glutamate dehydrogenase by lysosomal extracts (Timoneda, Wallace, and Grisolia, 1981), and ATP and other purine nucleotides stimulate the proteolytic inactivation of ornithine transcarbamylase (Navarro and Grisolia, 1984). Therefore, the interior milieu of the mitochondria may markedly influence the rates of degradation of various enzymes. It may explain why some isozymes present both in the cytosol and mitochondrial matrix have different turnover rates. For example, mitochondrial fumarase turns over considerably more slowly than the cytosolic isozyme in rat liver, although the two enzymes appear immunologically, enzymatically, and structurally similar (Kobayashi, Kamimura, and Tuboi, 1982). Alternatively, such results may be explained by artefactual differences in the extent of re-utilization of labeled precursors in mitochondria and in the cytosol or by differences in the specificity of the pathways responsible for their removal.

A different set of rules may govern the degradation of proteins embedded in membrane structures. There is evidence that the accessibility of membrane proteins to proteases may be an important factor. It has been shown that glucose repression in undividing yeast cells, which causes considerable alteration in the structure of the lipid membrane (i.e., higher ordering and lower mobility of hydrocarbon chains of constituting phospholipids), reduces the turnover of mitochondrial translation products (Luzikov et al., 1983). In isolated rat liver mitochondria, calcium-induced swelling of the organelle appears required for proteolysis (Rapoport, Dubiel, and Muller, 1983). In addition, increased susceptibility of swollen mitochondria to externally added proteases has been observed (Luzikov and Romanisha, 1972). Although only a few proteins have been studied, the orientation of the polypeptides in the membrane does not appear to be a crucial factor in determining their rates of degradation. Hare and Hodges (1982*b*) observed that the apocytochrome b and cytochrome b-c core subunit II, which are integral transmembrane proteins, are degraded faster than cytochrome c oxidase subunit III, which also spans the membrane. In addition, ATPase subunits facing the matrix side of the inner membrane turn over at the same rate as cytochrome c oxidase subunit II, which is exposed to the opposite side of the membrane (Hare and Hodges, 1982*b*).

Thus, a number of factors may influence the susceptibility of mitochondrial polypeptides for degradation and may account for some of the heterogeneity in their rates of degradation. In the final analysis, the most important determinant may prove to be the specificity of the degradative pathways.

ROLE OF CYTOPLASMIC PROTEOLYTIC PATHWAYS

A multiplicity of proteolytic pathways operate in animals cells, with distinct degradative systems that hydrolyze different classes of proteins or that become active only under certain physiological or pathological conditions. For example, lysosomes appear to participate in degradation of extracellular phagocytosed materials, of membrane proteins (e.g. acetylcholine receptors), and of certain long-lived proteins (Goldberg and St. John, 1976; Hershko and Ciechanover, 1982; Libby, Bersztajn, and Goldberg, 1980). Strong evidence indicates that lysosomes play a major role under conditions of nutritional deprivation and in the disposal of injured cytoplasmic components (Goldberg and St. John, 1976; Hershko and Ciechanover, 1982). By contrast, a soluble proteolytic system that is stimulated by ATP appears responsible for the breakdown of very short-lived proteins or those with abnormal structures, as may result from mutations, biosynthetic errors, or postsynthetic damages (Goldberg and St. John, 1976; Hershko and Ciechanover, 1982; Libby, Bursztajn, and Goldberg, 1980).

Numerous observations have led to the suggestion that mitochondria may be degraded within vesicular structures of lysosomes. The breakdown of mitochondrial proteins obeys first order kinetics (Lipsky and Pedersen, 1981*a;* Druyan, Debernard, and Rabinowitz, 1969; Russell, Burgess, and Mayer, 1982), seeming to indicate that the degradation of mitochondria is a random event. In other words, "new" mitochondria are as likely to be degraded as "old" ones and it has been suggested that the proximity of mitochondria to lysosomes, rather than the age, size, or quality, may be the primary factor in determining their rate of degradation (Lipsky and Pedersen, 1981*a*). It was shown that perinuclear distribution of micro-injected labeled mitochondria in hepatocyte monolayers precedes their degradation and that disruption of cytoskeletal elements (e.g., with colchicine), which prevents their distribution from the site of the injection, markedly reduces their degradation; thus, localization of mitochondria within the cell may be a very important factor in the degradation of mitochondrial proteins (Evans and Mayer, 1984). In addition, under certain conditions whole mitochondria or mitochondrial enzymes have been demonstrated within autophagic vacuoles (Knecht et al., 1980; Arstila, Shelburne, and Trump, 1972; Swift and Hruban, 1964). Breakdown of mitochondria and inactivation of mitochondrial enzymes are also observed when these organelles are incubated with lysosomes or lysosomal extracts at neutral pH (Navarro and Grisolia, 1984; Rubio and Grisolia, 1977; Allan and Welman, 1980; Grisolia et al., 1980). In hepatocytes, the degradation of certain outer mitochondrial membrane proteins and of polypeptides synthesized within the organelle is partially inhibited by chloroquine and methylamine (Evans and Mayer, 1982, 1984; Russell and Mayer, 1983). These weak bases accumulate within lysosomes and increase their pH sufficiently to prevent some of the action of acid proteases (Deduve, 1983). Chloroquine also inhibits lysosomal cathepsin B activity (Deduve, 1983). The degradation of mitochondrial translation products in hepatocytes is also inhibited 15 to 25% by the peptide aldehyde inhibitors, leupeptin and chymostatin (Evans and Mayer, 1983). Chymostatin inhibits cathepsin A, B, and D while leupeptin inhibits the thiol cathepsins (Umezawa, 1976) as well as the cytosolic neutral Ca^{++}-activated proteases (Waxman, 1981). Thus, a lysosomal route of degradation appears likely for at least some of the mitochondrial proteins. However, since mitochondrial proteolysis is only partially inhibited by agents that disrupt lysosomal function, proteolytic pathways other than the lysosome probably also participate in the breakdown of mitochondrial proteins.

Reticulocytes, which lack lysosomes, exhibit extensive degradation of mitochondria, ribosomes, and of many cellular proteins during maturation (Gasko and Danon, 1972; Rapoport, Dubiel, and Muller, 1981; Muller et al., 1980). In these cells, mitochondrial breakdown appears triggered by the action of a specific lipoxygenase that attacks the unsaturated fatty acids of the mitochondrial membranes, leading to lysis of the organelle (Schewe et al., 1975; Rapoport et al., 1979). Mitochondrial proteins then become substrates for a soluble ATP-dependent proteolytic pathway (Hershko and Ciechanover, 1982; Goldberg and St. John, 1976; Etlinger and Goldberg, 1977; Waxman et al., 1985). In support of this hypothesis, salicylhydroxamic acid, an inhibitor of lipoxygenase activity, inhibits markedly the degradation of mitochondrial proteins in reticulocytes (Rapoport, Dubiel, and Muller, 1981). It is still unknown whether this pathway for mitochondrial degradation is functioning in cells other than reticulocytes.

ATP-DEPENDENT PROTEOLYTIC PATHWAYS

The degradation of mitochondrial proteins within lysosomes or autophagic vacuoles or in the cell cytosol following lipoxygenase-induced lysis of the organelle is obviously indiscriminate. Although these proteolytic processes may be very important in mediating rapid changes in the cell content of mitochondria and in removing nonfunctioning or damaged organelles, they appear unlikely to participate in subtle regulation of enzyme levels within the mitochondria. They cannot account for the findings that the average half life of proteins differs in the several mitochondrial compartments and that within the same compartments different polypeptides and enzymes are degraded at distinct rates. It is probable that the accessibility of the outer membrane to direct attack by cytosolic proteolytic pathways may be responsible for its faster turnover rate when compared with the remainder of the mitochondria and may even explain the heterogeneity of turnover rates of certain outer membrane enzymes. By contrast, the sequence of events leading to heterogenous elimination of mitochondrial polypeptides embedded in the inner membrane or present within the matrix space has remained completely beyond our understanding. One possible mechanism is that such proteins may be selectively excreted by the mitochondria for degradation in the cytoplasm or lysosomes. This process appears unlikely as it would require very complex mechanisms (e.g., special signal sequences) similar to those necessary for protein entry into mitochondria (Neupert and Schatz, 1981). A more likely alternative is the existence within mitochondria of a proteolytic system that can selectively hydrolyze the short-lived proteins.

Early work showing large release of amino acids when isolated mitochondria were incubated at neutral pH suggested the presence of an endogenous proteolytic pathway (Alberti and Bartley, 1969; Ferdinand, Bartley, and Broomhead, 1973; Gear, Albert, and Bednarek, 1973; Subramanian et al., 1975), although such studies rarely excluded the presence of lysosomes or lysosomal enzymes in the isolated mitochondrial preparations. Mitochondria, however, clearly do have the capacity for limited proteolytic cleavage because many subunits of mitochondrial enzymes that are synthesized in the cytoplasm as large precursors undergo maturational cleavage in the mitochondria (Neupert and Schatz, 1981). It was also observed that inhibition of processing by mitochondria of the cytoplasmically synthesized enzyme carbamoyl phosphate synthase leads to the rapid degradation of the unprocessed polypeptide (Raymond and Shore, 1980). In addition, most proteins synthesized by isolated mitochondria, unlike those synthesized in vivo, appear to be degraded particularly rapidly. In rat liver and hepatoma cells, mitochondrial translation products have a mean half life of 50 to 70 hours (Hare and Hodges, 1982*b;* Evans and Mayer, 1983). By contrast, polypeptides synthesized by isolated yeast (Bakalkin et al., 1978; Kalnov et al., 1979) or liver mitochondria (Wheeldon and Lehninger, 1966; Wheeldon et al., 1974; Desautels and Goldberg, 1982*a*) have a half life of less than 60 minutes. In isolated liver mitochondria, newly synthesized polypeptides are degraded to amino acids that are then transported rapidly out of the mitochondria so no peptide intermediates accumulate in the organelle (Desautels and Goldberg, 1982*a*).

Inhibition of lysosomal activity with ammonium chloride or chloroquine or inhibition with leupeptin and chymostatin of the various proteases that contaminate the mitochondrial preparations does not affect the degradative process (Desautels and Goldberg, 1982*a*). Thus, neither intact lysosomes nor proteases released from such organelles participate in the rapid degradation of mitochondrial translation products in isolated mitochondria. Furthermore, the rate and extent of degradation of mitochondrial polypeptides are markedly increased when they are synthesized in the presence of low concentrations of puromycin, which induce their premature termination and release from ribosomes (Kalnov et al., 1979; Wheeldon et al., 1974; Desautels and Goldberg, 1982*a*). The extensive degradation of such incomplete polypeptides suggests that mitochondria, like intact animal and bacterial cells (Goldberg and St. John, 1976; Hershko and Ciechanover, 1982), have the capacity to selectively hydrolyze proteins with abnormal conformations. In the cells, such proteins may arise through mutations, biosynthetic errors, or postsynthetic damages such as those caused by oxidizing agents (Goldberg and Boches, 1982). Cytoplasmically encoded proteins with aberrant structures such as unprocessed polypeptides may also be substrate for the mitochondrial proteolytic pathway (Raymond and Shore, 1980). It appears likely that rapid elimination of polypeptides synthesized by isolated mitochondria may result from these newly made proteins being recognized as structurally abnormal. As most proteins synthesized by mitochondria associate with protein subunits synthesized in the cytoplasm, the lack of these subunits or of some other cofactors (e.g., heme (Saltzgaber-Muller and Schatz, 1978) or oxygen (Woodrow and Shatz, 1979) required for proper folding or assembly may induce their selective breakdown. In support of this hypothesis, instability of mitochondrial translation products has been shown in intact cells when cytoplasmic protein synthesis is prevented (Costantino and Attardi, 1977). In yeasts, respiratory chain components that do not participate in cell respiration are selectively eliminated (Galkin, Tsoi, and Luzikov, 1979*b*, 1980) and it has been suggested that mitochondrial proteolysis probably participate in an important way in mitochondrial biogenesis by selectively degrading nonfunctional components and insuring a balanced accumulation of the products of the nuclear and mitochondrial genome (Luzikov, Makhlis, and Galkin, 1976; Luzikov, 1981).

In bacterial and animal cells, many proteins are eliminated by a nonlysosomal pathway that is activated by ATP (Hershko and Ciechanover, 1982; Tanaka, Waxman, and Goldberg, 1983, 1984; Ciechanover, Finley, and Varshavsky, 1984). As in bacteria or in the cytosol of animal cells, the degradation of polypeptides synthesized by isolated yeast (Kalnov, 1979; Nascimento, 1983) or liver mitochondria (Wheeldon and Lehninger, 1966; Wheeldon et al., 1974; Desautels and Goldberg, 1982*a*) also requires ATP that can be either generated by oxidative phosphorylation or supplied externally (Desautels and Goldberg, 1982*a*). Cyanide, an inhibitor of respiration, or oligomycin and dinitrophenol, which uncouple oxidative phosphorylation, reduce proteolysis when mitochondria are provided with substrates for ATP generation (Kalnov et al., 1979; Desautels and Goldberg, 1982*a*). They do not affect proteolysis when exogenous ATP is provided but protein breakdown is then sensitive to atractyloside, an inhibitor of adenine nucleotide transport (Desautels and Goldberg, 1982*a*). Thus, for proteolysis, it is the nucleotide itself that is required, not the integrity of the electrochemical gradient across the inner mitochondrial membrane. Furthermore, it may also be inferred that the polypeptides synthesized in mitochondria, the ATP, and the proteolytic enzymes are most likely localized within the same space, enclosed by the inner mitochondrial membrane.

Not only is the presence of ATP required for intramitochondrial proteolysis, but its hydrolysis also appears essential. Vanadate, a potent inhibitor of various ATPases, blocks the degradation of mitochondrial translation products even in the presence of ATP and causes a marked stabilization of nearly all polypeptides newly synthesized in isolated liver mitochondria (Desautels and Goldberg, 1982*a*). Thus, ATP cleavage appears essential for an early event, probably the rate limiting step, in the degradation of mitochondrial translation products.

Kalnov and co-workers (1979) suggested that the breakdown of mitochondrial translation products in yeasts is probably due to the action of endogenous proteases associated with the mitochondrial inner membrane, since the degradation of these newly synthesized polypeptides could be observed in "inside-out" submitochondrial particles prepared by sonication and was inhibited by the addition of protease inhibitors such as leupeptin, antipain, chymostatin, and phenylmethyl sulfonylfluoride. However, the efficiency of the degradation of mitochondrial translation products was much lower than in intact mitochondria and it was not demonstrated that proteolysis in submitochondrial vesicles was still ATP-dependent.

In addition to an important role in the removal of proteins with abnormal conformations (e.g., incomplete polypeptides), evidence has been presented suggesting that ATP-dependent proteolytic pathways may be instrumental in mediating the constant renewal of mitochondrial proteins. Rapoport and co-workers reported ATP stimulation of lysine release from isolated liver mitochondria (Dubiel, Muller, and Rapoport, 1981; Rapport, Dubiel, and Muller, 1982). The release of amino acids from isolated mitochondria may be more representative of the bulk of protein turnover of the organelle than is the degradation of a select group of proteins synthesized on mitochondrial ribosomes in vitro (Rapoport, Dubiel, and Muller, 1982). Like the degradation of mitochondrial translation products, lysine liberation from isolated mitochondria is abolished by the addition of antimycin A, an inhibitor of mitochondrial respiration (Dubiel, Muller, and Rapoport, 1981; Rapoport, Dubiel, and Muller, 1982). In contrast to the degradation of mitochondrial translation products, which is inhibited when ATP supplied externally is prevented from entering the organelle, lysine liberation from isolated liver mitochondria is only slightly inhibited when carboxy atractylozide, a potent inhibitor of adenine nucleotide transport, is present simultaneously with ATP (Rapoport, Dubiel, and Muller, 1982). However, lysine liberation is markedly reduced when carboxy atractylozide prevents the exit of ATP produced within the organelle by oxidative phosphorylation (Rapoport, Dubiel, and Muller, 1982). These results suggest that an ATP-dependent proteolytic pathway may be located on the outer side of the inner mitochondrial membrane in contrast to the ATP-dependent degradation of mitochondrial translation products, which seems to occur within the space enclosed by the inner membrane (Desautels and Goldberg, 1982*a*). This disagreement about the localization of the ATP-stimulated proteolytic pathway(s) possibly reflects differences in the methods of measurement. For instance, caution should be exercised in interpreting results from amino acid release from isolated liver mitochondria because these preparations usually contain lysosomes whose low pH required for proteolytic activity is maintained by ATP-dependent proton pumps (Schneider, 1981). Alternatively, it is possible that two ATP-stimulated proteolytic pathways exist in mitochondria: one located in the intermembrane space, possibly bound to the outer surface of the inner membrane, and one localized within the matrix space. This interpretation would be consistent with the heterogenous rates of protein degradation in the two separate compartments.

ATP-DEPENDENT PROTEASE IN LIVER MITOCHONDRIA

A requirement for energy for proteolysis is not unique to mitochondria but has been originally described in bacteria (Goldberg and St. John, 1976) and in the cytosol of animal cells (Hershko and Ciechanover, 1982; Etlinger and Goldberg, 1977). In addition, ATP-dependent proteolysis has been described recently in another membrane-enclosed organelle, the chloroplast (Liu and Jagendorf, 1984; Malek et al., 1984). In reticulocytes, ATP appears to serve two distinct roles. Hershko and Ciechanover (1982) have presented extensive evidence that ATP is required for an initial recognition reaction in which the proteins to be degraded are covalently modified ("tagged") by ATP-dependent ligation of a small polypeptide, ubiquitin, to enhance their susceptibility to hydrolysis by cytosolic proteases. In addition, an ATP-activated process that is independent of ubiquitin conjugation also appears to play an essential role in protein breakdown. This second ATP-dependent step, which is sensitive to vanadate and is able to utilize nucleotides other than ATP (unlike the ubiquitin-activated step), appear to require direct interaction of the nucleotide with proteolytic enzymes (Tanaka, Waxman, and Goldberg, 1983, 1984). Reticulocytes and other mammalian cells contain a soluble alkaline endoprotease that is stimulated by ATP (DeMartino and Goldberg, 1979; Waxman et al., 1985).

In *Eschericia coli,* the energy requirement for protein breakdown results solely from the ATP-requirement for Protease La, the product of the lon gene (Chung and Goldberg, 1981; Charette, Henderson, and Markovitz, 1981), which catalyzes the initial endoproteolytic steps in the degradation of proteins (Kowit and Goldberg, 1977; Gottesman and Zipser, 1978). The initial cleavages by Protease La are followed by the action of numerous

ATP-independent proteases that complete the degradative process (Swamy and Goldberg, 1981). Protease La is a very unique enzyme. It was originally purified as a DNA-binding protein (Zehnbauer et al., 1981) and is normally a tetramer composed of identical 94000-dalton subunits (Chung and Goldberg, 1981; Zehnbauer et al., 1981); also, it hydrolyzes both ATP and protein in a linked fashion (Waxman and Goldberg, 1982) and its ATPase activity is stimulated by proteins that are substrates for the protease (Charette, Henderson, and Markovitz, 1981; Waxman and Goldberg, 1982).

As in bacteria and in the cytosol of animal cells, rat liver mitochondria contain an ATP-dependent protease that can account for the energy requirement for proteolysis in the intact mitochondria (Desautels and Goldberg, 1982*b*). Soluble extracts of liver mitochondria show significant hydrolysis of labeled casein (Desautels and Goldberg, 1982*b*) and other large proteins that is stimulated twofold to sevenfold by ATP (Desautels and Goldberg, 1982*b;* Desautels and Goldberg, 1985). When submitochondrial fractions are isolated, the ATP-stimulated proteolytic activity is localized to the matrix, the same space in which an ATP-dependent proteolytic pathway is observed (Desautels and Goldberg, 1982*b*). There is also a small stimulation (40%) by ATP of casein hydrolysis in the intermembrane space fraction that may be due to the breakage of the organelle during fractionation (Desautels and Goldberg, 1982*b*). Alternatively, this activity may be due to the presence of a second ATP-dependent proteolytic pathway (Rapoport, Dubiel, and Muller, 1982). No ATP-dependent proteolysis is found associated with the mitochondrial membranes (Desautels and Goldberg, 1982*b*). Extensive purification of the ATP-stimulated proteolytic activity of mitochondrial soluble extracts revealed the presence of a unique and unusually large serine protease (MW 550000) that is active at alkaline pH but only in the presence of both magnesium and ATP (Desautels and Goldberg, 1982*b*). This enzyme does not appear to be a lysosomal or cytoplasmic contaminant. It is isolated from mitochondria treated with digitonin, which removed 85 to 98% of all lysosomal enzymes contaminating the preparation. It has no activity at pH values below 7, where lysosomal enzymes are most active, and it is not affected by inhibitors of lysosomal proteases (Desautels and Goldberg, 1982*b*). Furthermore, its molecular weight (550000) is greater than any lysosomal protease. Its properties also clearly distinguish the mitochondrial enzyme from the ATP-stimulated protease reported in liver cytosol (DeMartino and Goldberg, 1979), but resemble those of a soluble ATP-requiring protease from murine erythroleukemia cells (Waxman et al., 1985).

Several observations indicate that the mitochondrial ATP-dependent protease is probably involved in the breakdown of labile polypeptides synthesized by isolated liver mitochondria. As in the intact organelle, hydrolysis of ATP is required for enzyme activity. Nonhydrolyzable ATP analogs cannot replace ATP, and inhibitors of ATPases such as vanadate and quercetin block its activity (Desautels and Goldberg, 1982*b*). In fact, vanadate-sensitive ATPase activity co-purifies with the proteolytic activity and both processes show a similar Km for ATP (Desautels and Goldberg, 1982*c;* 1985). In addition, the isolated protease is inhibited by the same concentrations of vanadate that block protein breakdown in intact mitochondria (Desautels and Goldberg, 1982*b*). Furthermore, in accord with the observation that ATP hydrolysis is apparently required for some initial steps in the degradation of mitochondrial proteins, the ATP-dependent protease is an endoprotease capable of degrading large proteins (e.g., casein) to polypeptides with molecular weights greater than 1500 (Desautels and Goldberg, 1982*b,* 1982*c*). Thus, the ATP-dependent protease is probably responsible for the initial cleavages of proteins to be degraded within mitochondria.

A number of important questions remain to be answered concerning the physiological role(s), the specificity, and the mechanism of action of the mitochondrial protease. Although evidence has been presented that indicates a role for the enzyme in the breakdown of abnormal proteins, it is unclear whether the protease can also hydrolyze short-lived regulatory enzymes and membrane proteins or could be involved in the processing of cytoplasmically made precursors of mitochondrial proteins. In other words, its quantitative contribution to the constant renewal of mitochondrial proteins in cells under basal conditions has not yet been established. It is also unknown whether the enzyme is involved in the accelerated degradation of mitochondria under certain physiological or pathological conditions in which the cellular content of mitochondrial proteins is altered.

There are two crucial questions concerning the mechanism of this novel protease: How does the enzyme recognize what is a protein with an abnormal structure, and what is the role of ATP in its functioning? Recent work by Goldberg and Waxman (1985) indicates that ATP hydrolysis by protease La, the bacterial ATP-dependent protease, is not required for peptide bond cleavage or modification of the protein substrates, but is required for temporary inactivation of the enzyme. Their model involves a cyclical series of reactions in which the binding of ATP and protein substrates allosterically activates the protease and allows peptide bond cleavage. The hydrolysis of ATP is a later event that leads to inactivation of the enzyme until additional ATP's are bound. Goldberg and Waxman's model is based on the observation that in contrast to the degradation of large proteins, which requires concomittant ATP hydrolysis (Waxman and Goldberg, 1982), binding of Mg^{2+}-ATP alone is sufficient to allow cleavage of specific fluorogenic tetrapeptides such as glutaryl-ALA-ALA-Phe-methoxynaphtylamine.

The mitochondrial protease has a number of unique structural and functional properties that closely resemble those of protease La suggesting that their mechanism of action may be very similar. Both enzymes are unusually large proteases (Chung and Goldberg, 1981; Charette, Henderson, and Markovitz, 1981; Desautels and Goldberg, 1982*b*), and it has been recently suggested that both their size and their utilization of ATP may underlie the selectivity and regulation of intracellular proteolysis (Waxman and Goldberg, 1982). Although the subunit composition of the mitochondrial protease is still undefined, it differs from that of protease La in that there is no 94000 dalton polypeptide in the mitochondrial preparations (Desautels and Goldberg, 1982*c*). Both protease La and the mitochondrial enzyme are serine endoproteases that require ATP and divalent cations for protein degradation, have a similar Km for ATP, and are inhibited by similar concentrations of vanadate (Waxman and Goldberg, 1982; Desautels and Goldberg, 1982*b*). For both enzymes, protein cleavage appears coupled to ATP hydrolysis (Waxman and Goldberg, 1982; Desautels and Goldberg, 1985; Desautels and Goldberg, 1982*c*). Both processes have the same Km for ATP, and inhibitors of the ATPase activity of these enzymes also block their proteolytic activity (Waxman and Goldberg, 1982; Desautels and Goldberg, 1982*b;* Desautels, Waxman and Goldberg, 1986; Desautels and Goldberg, 1982*c*). In addition, casein, the protein used as substrate, activates both enzymes and markedly stimulates ATP hydrolysis. For the bacterial enzyme, it was clearly shown that only proteins that are substrates for the protease can activate the enzyme (Waxman and Goldberg, 1982), and it has been suggested that induction of the proteolytic activity of protease La and of the mitochondrial protease by protein substrates probably ensures that these enzymes do not hydrolyze proteins in a nonselective way (Waxman and Goldberg, 1982; Desautels and Goldberg, 1982*c*). Regulation of their ATPase activities by protein substrates may also be important to avoid depletion of the cell's or the organelle's high-energy phosphate stores. Furthermore, the ATP-dependent mitochondrial protease cleaves the same fluorogenic peptides recognized by protease La (Desautels and Goldberg, 1985). Therefore it appears likely that the mechanism of action of the mitochondrial enzyme as well as the recognition of what constitutes an abnormal protein may have been conserved during the evolution of mitochondria from bacteria that probably originally existed in symbiotic association with eukaryotic cells (Desautels and Goldberg, 1982*c*).

ATP-INDEPENDENT PROTEASES

Since mitochondria have the capacity to degrade proteins to their constituent amino acids, there must be a variety of endoproteases and peptidases associated with the organelle in addition to the ATP-dependent protease. Several reports have described the presence of proteases in mitochondria of yeasts and mammalian cells. Soluble extracts of liver mitochondria contain significant proteolytic activity at neutral or alkaline pH against casein (Desautels and Goldberg, 1982*b;* Duque-Magalhaes, 1979; Duque-Magalhaes and Regnier, 1982), insulin (Desautels and Goldberg, 1982*b*), or cytochrome c (Duque-Magalhaes, 1979; Duque-Magalhaes and Menezes-Ferreira, 1980). Differences in pH optimum and inhibitor sensitivity indicate the presence of several distinct enzymes. Recently it was reported that liver mitochondria also contain calcium-activated neutral proteolytic activity with possibly two such enzymes—one activated by low calcium concentration ($< 100\,\mu$M) and one active only at high concentration (1mM) (Beer et al., 1982). Liver mitochondrial membrane extracts also show proteolytic activity against casein (Desautels and Goldberg, 1982*b*), insulin (Desautels and Goldberg, 1982*b;* Hare, 1978), and cytochrome c (Duque-Magalhaes and Menezes-Ferreira, 1980). In addition, a methionine aminopeptidase associated with both liver mitochondria and microsomes has been reported (Freitas et al., 1981). In yeasts, the mitochondrial membranes possess at least two esterases and two proteases that digest cytochrome (Novikova, Zubatov, and Luzikov, 1981). It has been suggested that these enzymes may be involved in the cleavage of mitochondrial translation products because they are sensitive to the same inhibitors (leupeptin, antipain, chymostastin, and phenylmethylsulfonylfluoride) that block mitochondrial proteolysis in submitochondrial particles (Kalnov et al., 1979; Freitas et al., 1981).

However, in contrast to the ATP-dependent protease whose properties closely coincide with the characteristics of the proteolytic pathway in the intact organelle, it is sometimes very difficult to ascribe a specific function to proteases isolated from mitochondrial fractions or even to unambiguously identify some of these enzymes as being truly of mitochondrial origin. One problem is the unspecificity of the protein substrate utilized. Few of the proteases so far described have been purified and their natural substrates are unknown. A more difficult problem is the presence of lysosomes or of protease-rich organelles in the mitochondrial preparations in addition to unspecific adsorption of cytosolic or lysosomal enzymes to the mitochondrial membranes that occurs during tissue homogenization or cell rupture. To minimize such a problem, mitochondria are usually treated with low concentrations of digitonin, which lyzes the lysosomes preferentially. Such treatment markedly reduces the release of amino acids when mitochondria are incubated at neutral pH (Rubio and Grisolia, 1977; Manjunath, Padmanaban, and Cama, 1979) and reduces the capacity of mitochondrial extracts to hydrolyze casein (Duque-Magalhaes, 1979). By contrast, cytochrome c-degrading activity is unchanged in digitonin-washed liver mitochondria and it was shown that the two mitochondrial proteases able to digest cytochrome c are not found in isolated lysosomes (Duque-Magalhaes and Menezes-

Ferreira, 1980). However, even a protease isolated from extensively washed, digitonin-treated mitochondria may not be truly of mitochondrial origin. For instance, group-specific proteases that inactivate specifically the apo-forms of certain pyridoxal-phosphate enzymes were purified from the mitochondrial fraction of bone marrow cells, liver, skeletal muscle, and intestine (Aoki et al., 1975; Aoki, 1978; Katunuma et al., 1975; Banno, Morris, and Katunuma, 1978). These enzymes, of great interest because of their capacity to degrade δ-amino levulinate synthase, a short-lived mitochondrial matrix enzyme, were later shown to reside within mast cell granules whose sedimentation properties resemble those of mitochondria (Woodbury, Gruzenski, and Lagunoff, 1978; Haas, Heinrich, and Sasse, 1979). Likewise, an histone-degrading protease and a carboxypeptidase originally purified from liver mitochondrial inner membranes (Haas and Heinrich, 1978, 1979) were also of mast cell origin (Haas, Heinrich, and Sasse, 1979). The physiochemical and immunological properties of the histone-degrading enzyme were identical to those of the chymotrypsin-like protease of peritoneal mast cells, and the recovery of both the protease and carboxypeptidase was markedly reduced when animals were injected with compound 48/80, known to cause degranulation of mast cells (Haas, Heinrich, and Sasse, 1979).

Despite these difficulties, not all proteolytic activity associated with mitochondria is of lysosomal or mast cell origin (Dean, 1983). Digitonin-washed liver mitochondria isolated from animals injected with compound 48/80 still retain a significant capacity to hydrolyze proteins at alkaline pH, as do mitochondria isolated from brain, which have few or no mast cells (Duque-Magalhaes and Regnier, 1982). Finally, there is a group of mitochondrial proteases whose properties, like the ATP-dependent protease, closely match known proteolytic requirements in the intact organelle and therefore are much less likely to be cytoplasmic, lysosomal, or mast cell contaminants.

PROCESSING OF PRECURSOR PROTEINS

Recent studies indicate that most mitochondrial proteins are synthesized as large precursors with signal peptides of MW 2000 to 8000 (Neupert and Schatz, 1981). These proteins are then transported across the organelle membranes and processed to their mature form. Import of cytoplasmically made mitochondrial proteins is energy-dependent (Neupert and Schatz, 1981; Gasser, Daum, and Schatz, 1982). An electrochemical gradient across the mitochondrial inner membrane but not ATP itself is required for protein import (Gasser, Daum, and Schatz, 1982). The precursors are then processed to the mature polypeptides by soluble proteases located in the mitochondrial matrix. Several independent reports describe proteases that cleave the precursors of cytochrome c oxidase subunit V (Bohni, Daum, and Schatz, 1983), ornithine carbamoyl transferase (Mori et al., 1980; Conboy, Fenton, and Rosenberg, 1982; Miura et al., 1982), and F_1-ATPase subunits (McAda and Douglas, 1981). All processing enzymes characterized from either yeasts or liver mitochondria have numerous features in common. All are soluble enzymes located in the mitochondrial matrix and all have a pH optimum of 7.5. Their molecular weight is approximately 108000 to 115000 (Miura et al., 1982; McAda and Douglas, 1981). They are metalloenzymes that are inhibited by metal chelators such as EDTA and o-phenantroline, and they are unaffected by seryl protease inhibitors (Bohni, Daum, and Schatz, 1983; Mori et al., 1980; Conboy, Fenton, and Rosenberg, 1982; Miura et al., 1982; McAda and Douglas, 1981). These proteases appear quite specific for precursors of mitochondrial proteins. They do not attack mature enzymes or nonmitochondrial proteins (Bohni, Daum, and Schatz, 1983; Miura et al., 1982). It appears likely that the mitochondrial-processing enzymes are common to several protein precursors (McAda and Douglas, 1981). For example, the protease that cleaves the precursor of ornithine carbamoyl transferase is also found in mitochondria from kidney, spleen, heart, and ascite tumor cells, all of which lack ornithine carbamoyl transferase (Miura et al., 1982). It appears also likely that there is more than one processing enzyme. Miura et al. (1982) found that the protease that cleaves the precursor of ornithine carbamoyl transferase and of serine-pyruvate amino transferase, another matrix enzyme, does not process the precursor of carbamoyl phosphate synthetase I. In addition to the processing enzyme, other proteases and peptidases, even possibly the ATP-dependent protease, are probably involved in the elimination of the signal peptides that otherwise might accumulate in the organelle.

SUMMARY AND PERSPECTIVES

The maintenance of a stable and functional mitochondrial mass in the cell requires constant insertion and removal of proteins from the organelles. There is a marked heterogeneity in the turnover rates of mitochondrial proteins, and the mechanisms underlying such heterogeneity are probably crucial for allowing rapid changes in mitochondrial enzyme composition in response to various physiological stimuli. The question of how mitochondrial proteins can be removed from the organelles at heterogenous rates is not easily answered. Several proteolytic pathways appear to be involved in the breakdown of mitochondrial proteins. Which pathway is involved probably depends largely on the localization of the proteins in the mitochondria. Outer membrane proteins, which are accessible from the cell cytosol, are probably degraded by proteolytic pathways exogenous to the mitochondria.

The exact sequence of events is unclear, but there may be some involvement of the lysosomes in their degradation. By contrast, some of the heterogeneity of turnover rates of proteins embedded in the inner membrane or localized within the matrix may be best explained by proteolytic action within the organelle. In addition, intramitochondrial proteolysis appears to be required for inactivation of short-lived regulatory enzymes, for the degradation of proteins with abnormal conformations, for the processing of cytoplasmically synthesized precursors of mitochondrial enzymes, and for the elimination of the signal peptides. To fulfill such a multiplicity of functions, mitochondria contain a variety of proteases and a proteolytic pathway capable of degrading proteins to their constitutent amino acids. As in bacteria or in the cytosol of animal cells, this proteolytic pathway requires ATP. Hydrolysis of the nucleotide appears necessary for some initial endoproteolytic cleavages of labile mitochondrial proteins. As in bacteria, the energy requirement for intramitochondrial proteolysis results from the requirement for ATP of a serine endoprotease whose structural and functional characteristics resemble closely those of the ATP-dependent protease of *Eschericia coli,* protease La. There is evidence that the mitochondrial protease, like protease La, is involved in the breakdown of abnormal proteins, but its contribution to the degradation of short-lived regulatory enzymes as well as to the overall turnover of mitochondrial proteins remains to be established. There are numerous other proteases in mitochondria, some of which show exquisite selectivity, such as the proteases that cleave the signal peptides required for entry into the organelle of the precursors of numerous cytoplasmically made mitochondrial proteins.

Superimposed onto the turnover of proteins within individual mitochondria is the regulation of the cell content of organelles. The neuro-endocrine signals and the mechanisms responsible for altering the balance between degradation and replacement of whole mitochondria may vary in different organs and have been very poorly defined. They may involve lysosomal engulfment or lysis of the organelle followed by protein breakdown in the cell cytosol. The possible contribution of the mitochondrial proteolytic pathway in accelerated mitochondrial degradation is unknown.

ACKNOWLEDGMENTS

I would like to thank Dr. A. L. Goldberg for his encouragement and support. The research outlined in this review was carried out in his laboratory. I am also extremely grateful to Mrs. J. Kripki for her expert assistance in the preparation of this manuscript and to the Medical Research Council of Canada for financial support.

REFERENCES

Alberti, K. G. M. M., and W. Bartley, 1969,The localization of proteolytic activity in rat liver mitochondria and its relation to mitochondrial swelling and aging, *Biochem. J.* **111:**763-776.

Albin, R., R. T. Dowell, R. Zak, and M. Rabinowitz, 1973, Synthesis and degradation of mitochondrial component as in hypertrophied rat heart, *Biochem. J.* **136:**629-637.

Allan, R. M., and E. Welman, 1980, Proteolysis of isolated mitochondria by myocardial lysosomal enzymes, *Biochem. J.* **190:**139-144.

Aoki, Y., 1978, Crystallization and characterization of a new protease in mitochondria of bone marrow cells, *J. Biol. Chem.* **253:**2026-2032.

Aoki, Y., S. Muranaka, K. Nakabayashi, and Y. Ueda, 1979, δ-aminolevulinic acid synthetase in erythroblasts of patients with pyridoxine-responsive anemia, *J. Clin. Invest.* **64:**1196-1203.

Aoki, Y., G. Urata, F. Takaku, and N. Katunuma, 1975, A new protease inactivating δ-aminolevulinic acid synthetase in mitochondria of human bone marrow cells, *Biochem. Biophys. Res. Commun.* **65**(2):567-574.

Arstila, A. V., J. D. Shelburne, and B. F. Trump, 1972, Studies on cellular autophagocytosis. A histochemical study on sequential alterations of mitochondria in the glucagon-induced autophagic vacuoles of rat liver, *Lab. Invest.* **27:**317-327.

Aschenbrenner, V., R. Druyan, R. Albin, and M. Rabinowitz, 1970, Haem a, cytochrome c and total protein turnover in mitochondria from rat heart and liver, *Biochem. J.* **119:**157-160.

Aschenbrenner, V., R. Zak, A. F. Cutilletta, and M. Rabinowitz, 1971, Effect on hypoxia on degradation of mitochondrial components in rat cardiac muscle, *Am. J. Physiol.* **221:**1418-1425.

Bakalkin, G. Y., S. L. Kalnov, A. V. Galkin, A. S. Zubatov, and V. N. Luzikov, 1978, The lability of the products of mitochondrial protein synthesis in saccharomyces cerevisiae, *Biochem. J.* **170:**569-576.

Banno, Y., H. P. Morris, and N. Katunuma, 1978, The serine protease from rat liver and hepatoma 8999, *J. Biochem.* **83:**1545-1554.

Beer, D. G., J. J. Hjelle, D. R. Petersen, and A. M Malkinson, 1982, Calcium-activated proteolytic activity in rat liver mitochondria, *Biochem. Biophys. Res. Commun.* **109:**1276-1283.

Bohni, P. C., G. Daum, and G. Schatz, 1983, Import of proteins into mitochondria, *J. Biol. Chem.* **258:**4937-4943.

Bukowiecki, L., and J. Himms-Hagen, 1971, Decreased half-life of some mitochondrial proteins in skeletal muscle and brown adipose tissue of cold-acclimated rats. *Can. J. Physiol. Pharmacol.* **49:**1015-1018.

Burgess, R. J., J. H. Walker, and R. J. Mayer, 1978, Choice of precursors for the measurement of protein turnover by the double-isotope method, *Biochem. J.* **176:**919-926.

Charette, M. E., G. W. Henderson, and A. Markovitz, 1981, ATP-hydrolysis-dependent protease activity of the lon (cap R) protein of *Eschericia coli, Natl. Acad. Sci. (USA) Proc.* **78:**4728-4732.

Chee, P. Y., and R. W. Swick, 1976, Effect of dietary protein and Tryptophan on the turnover of rat liver ornithine aminotransferase, *J. Biol. Chem.* **251:**1029-1034.

Chung, C. H., and A. L. Goldberg, 1981, The product of the lon (cap R) gene in *Eschericia coli* is the ATP-dependent Protease La, *Natl. Acad. Sci. (USA) Proc.* **78:**4931-4935.

Ciechanover, A., D. Finley, and A. Varshavsky, 1984, Ubiquitin dependence of selective protein degradation demonstrated in the mammalian cell cycle mutant ts85, *Cell* **37:**57-66.

Conboy, J. G., W. A. Fenton and L. E. Rosenberg, 1982, Processing of pre-ornithine transcarbamylase requires a zinc-dependent protease localized to the mitochondrial matrix, *Biochem. Biophys. Res. Commun.* **105:**1-7.

Costantino, P., and G. Attardi, 1977, Metabolic properties of the products of mitochondrial protein synthesis in Hela cells, *J. Biol. Chem.* **252:**1702-1711.

Dean, B., 1983, Proteolysis in mitochondrial preparations and in lysosomal preparations derived from rat liver, *Arch. Biochem. Biophys.* **227:**154-163.

DeMartino, G. N., and A. L. Goldberg, 1979, Identification and partial purification of an ATP-stimulated alkaline protease in rat liver, *J. Biol. Chem.* **254:**3712-3715.

Desautels, M., and J. Himms-Hagen, 1980, Parallel regression of cold-induced changes in ultrastructure, composition and properties of brown adipose tissue mitochondria during recovery of rats from acclimation to cold, *Can. J. Biochem.* **58:**1057-1068.

Desautels, M., and A. L. Goldberg, 1982*a*, Liver mitochondria contain an ATP-dependent, vanadate-sensitive pathway for the degradation of proteins, *Natl. Acad. Sci.(USA) Proc.* **79:**1869-1873.

Desautels, M., and A. L. Goldberg, 1982*b*, Demonstration of an ATP-dependent, vanadate-sensitive endoprotease in the matrix of rat liver mitochondria, *J. Biol. Chem.* **257:**11673-11679.

Desautels, M., and A. L. Goldberg, 1982*c*, Rat liver mitochondria contain an ATP-dependent pathway for protein breakdown and an ATP-dependent endoprotease, *Fed. Proc.* **41:**866.

Desautels, M., and A. L. Goldberg, 1985, The ATP-dependent breakdown of proteins in mammalian mitochondria, *Biochem. Soc. Trans.* **13:**290-293.

Desautels, M., and H. M. Yuen, 1984, Loss of mitochondria is a primary response to food deprivation in mouse brown adipose tissue, *Fed. Proc.* **43:**1063.

Dice, J. F., and A. L. Goldberg, 1975, Relationship between in vivo degradative rates and isoelectric points of proteins, *Natl. Acad. Sci. (U.S.A.) Proc.* **72:**3893-3897.

Druyan, R., B. Debernard, and M. Rabinowitz, 1969, Turnover of cytochromes labeled with δ-aminolevulinic acid-^{3}H in rat liver, *J. Biol. Chem.* **244:**5874-5878.

Dubiel, W., M. Muller, and S. Rapoport, 1981, ATP-dependent proteolysis of rat liver mitochondria, *Acta Biol. Med. Germ.* **40:**K25-K29.

Duque-Magalhaes, M. C., 1979, On a Neutral Proteolytic system in rat liver mitochondria, *FEBS Lett.* **105:**(2):317-320.

Duque-Magalhaes, M. C., and M. M. Menezes-Ferreira, 1980, Cytochrome c degrading activity in rat liver mitochondria, *Biochem. Biophys. Res. Commun.* **93:**106-112.

Duque-Magalhaes, M. C., P. Regnier, 1982, Study on the localization of proteases of mitochondrial origin, *Biochimie* **64:**907-913.

Ernster, L., and G. Schatz, 1981, Mitochondria: A historical review, *J. Cell Biol.* **91**(3):227s-255s.

Etlinger, J. D., and A. L. Goldberg, 1977, A soluble ATP-dependent proteolytic system responsible for the degradation of abnormal proteins in reticulocytes, *Natl. Acad. Sci. (USA) Proc.* **74:**54-58.

Evans, P. J., and R. J. Mayer, 1982, Organelle membrane-cell fusion: destruction of transplanted mitochondrial proteins in hepatocyte monolayers, *Biochem. Biophys. Res. Commun.* **107:**51-58.

Evans, P. J., and R. J. Mayer, 1983, Degradation of transplanted mitochondrial proteins by hepatocyte monolayers, *Biochem. J.* **216:**151-161.

Evans, P. J., and R. J. Mayer, 1984, Comparison of the degradative fate of monoamine oxidase in endogenous and transplanted mitochondrial outer membrane in rat hepatocytes, *Biochem. J.* **219:**61-72.

Ferdinand, W., W. Bartley, and V. Broomhead, 1973, Amino acid production in isolated rat liver mitochondria, *Biochem. J.* **134:**431-436.

Freitas, J. O., Jr., C. Termignoni, D. R. Borges, C. A. M. Sampaio, J. L. Prado, and J. A. Guimaraes, 1981, Methionine aminopeptidase associated with liver mitochondria and microsomes, *Int. J. Biochem.* **13:**991-997.

Galkin, A. V., T. V. Tsoi, and V. N. Luzikov, 1979*a*, Abnormalities in mitochondrial respiratory chain assembly and their proteolytic elimination, *FEBS Lett.* **103:**111-113.

Galkin, A. V., T. V. Tsoi, and V. N. Luzikov, 1979*b*, Regulation of mitochondrial biogenesis: Further evidence for proteinase involvement, *FEBS Lett.* **105:**373-375.

Galkin, A.V., T. V.Tsoi, and V. N. Luzikov, 1980, Regulation of mitochondrial biogenesis, *Biochem. J.* **190:**145-156.

Gasko, O., and D. Danon, 1972, Deterioration and disappearance of mitochondria during reticulocyte maturation, *Exp. Cell Res.* **75:**159-169.

Gasser, S. M., G. Daum, and G. Schatz, 1982, Import of proteins into mitochondria, *J. Biol. Chem.* **257**:13034-13041.

Gear, A. R. L., 1970, Inner- and outer-membrane enzymes of mitochondria during liver regeneration, *Biochem. J.* **120**:577-587.

Gear, A. R. L., A. D. Albert, and J. M. Bednarek, 1973, The effect of the hypocholesterolemic drug clofibrate on liver mitochondrial biogenesis, *J. Biol. Chem.* **249**:6495-6504.

Glass, R. D., and D. Doyle, 1972, On the measurement of protein turnover in animal cells, *J. Biol. Chem.* **247**:5234-5242.

Goldberg, A. L., and F. S. Boches, 1982, Oxidized proteins in erythrocytes are rapidly degraded by the ATP-dependent proteolytic system, *Science* **215**:1107-1109.

Goldberg, A. L., and J. F. Dice, 1974, Intracellular protein degradation in mammalian and bacterial cells, *Ann. Rev. Biochem.* **43**:835-869.

Goldberg, A. L., and A. C. St. John, 1976, Intracellular protein degradation in mammalian and bacterial cells, *Ann. Rev. Biochem.* **45**:747-803.

Goldberg, A. L., and L. Waxman, 1985, The role of ATP hydrolysis in the breakdown of proteins and peptides by protease La from Escherichia coli, *J. Biol. Chem.* **260**:12029-12034.

Gottesman, S., and D. Zipser, 1978, Deg phenotype of *Eschericia coli* lon mutants, *J. Bacteriol.* **133**:844-851.

Grisolia, S., E. Knecht, J. Hernandez-Yago, and R. Wallace, 1980, Turnover and degradation of mitochondria and their proteins, in *Protein Degradation in Health and Disease,* CIBA Foundation Symposium, vol. 75, pp. 167-188.

Gross, N. J., G. S. Getz, and M. Rabinowitz, 1969, Apparent turnover of mitochondrial deoxyribonucleic acid and mitochondrial phospholipids in the tissues of the rat, *J. Biol. Chem.* **244**:1552-1562.

Haas, R., and P. C. Heinrich, 1978, The localization of an intracellular membrane-bound proteinase from rat liver, *Eur. J. Biochem.* **91**:171-178.

Haas, R., and P. C. Heinrich, 1979, Novel SH-type carboxypeptidase in the inner membrane of rat-liver mitochondria, *Eur. J. Biochem.* **95**:9-15.

Haas, R., P. C. Heinrich, and D. Sasse, 1979, Proteolytic enzymes of rat liver mitochondria. Evidence for a mast cell origin, *FEBS Lett.* **103**:168-171.

Hare, J. F., 1978, A novel proteinase associated with mitochondrial membranes, *Biochem. Biophys. Res. Commun.* **83**:1206-1215.

Hare, J. F., and R. Hodges, 1982*a,* Turnover of mitochondrial matrix polypeptides in hepatoma monolayer cultures, *J. Biol. Chem.* **257**:12950-12953.

Hare, J. F., and R. Hodges, 1982*b,* Turnover of mitochondrial inner membrane proteins in hepatoma monolayer cultures, *J. Biol. Chem.* **257**:3575-3580.

Hershko, A., and A. Ciechanover, 1982, Mechanisms of intracellular protein breakdown, *Ann. Rev. Biochem.* **51**:335-364.

Himms-Hagen, J., E. Dittmar, and G. Zaror-Behrens, 1980, Polypeptide turnover in brown adipose tissue mitochondria during acclimation of rats to cold, *Can. J. Biochem.* **58**:336-344.

Ip, M. M., P. Y. Chee, and R. W. Swick, 1974, Turnover of hepatic mitochondrial ornithine aminotransferase and cytochrome oxidase using [^{14}C] carbonate as tracer, *Biochim. Biophys. Acta* **354**:29-38.

Kalnov, S. L., L. A. Novikova, A. S. Zubatov, and V. N. Luzikov, 1979, Proteolysis of the products of mitochondrial protein synthesis in yeast mitochondria and submiochondrial particles, *Biochem. J.* **182**:195-202.

Katunuma, N., E. Kominami, K. Kobayashi, Y. Banno, K. C. Suzuki, Y. Hamaguchi, K. Chichibu, and T. Katsunuma, 1975, Studies of new intracellular proteases in various organs of rat, *Eur. J. Biochem.* **52**:37-50.

Knecht, E., J. Hernandez-Yago, A. Martinez-Ramon, and S. Grisolia, 1980, Fate of proteins synthesized in mitochondria of cultured mammalian cells revealed by electron microscope radioautography, *Exp. Cell Res.* **125:**191-199.

Kobayashi, K., T. Manimura, and S. Tuboi, 1982, Studies on the turnover rates of cytosolic and mitochondrial fumarases in rat liver, *J. Biochem.* **91:**1511-1518.

Kowit, J. D., and A. L. Goldberg, 1977, Intermediate steps in the degradation of a specific abnormal protein in *Eschericia coli, J. Biol. Chem.* **52:**8350-8357.

Libby, P., S. Bursztajn, and A. L. Goldberg, 1980, Degradation of the acetylcholine receptor in cultured muscle cells: selective inhibitors and the fate of undegraded receptors, *Cell* **19:**481-491.

Lipsky, N. G., and P. L. Pedersen, 1981*a*, Mitochondrial turnover in animal cells, *J. Biol. Chem.* **256:**8652-8657.

Lipsky, N. G., and P. L. Pedersen, 1981*b*, Perturbation by clofibrate of mitochondrial levels in animal cells. Implications for a model of mitochondrial genesis, *J. Biol. Chem.* **257:**1473-1481.

Liu, X. Q., and A. T. Jagendorf, 1984, ATP-dependent proteolysis in pea chloroplasts, *FEBS Lett.* **166:**248-251.

Lovaas, E., 1974, Evidence for a proteolytic system in rat liver mitochondria, *FEBS Lett.* **45:**244-247.

Luzikov, V. N., 1981, Control over assembly of the mitochondrial membrane: Selection by a performance criterion, *FEBS Lett.* **125:**131-133.

Luzikov, V. N., T. A. Makhlis, and A. V. Galkin, 1976, A probable role of cell proteinases in the biogenesis of mitochondria in yeast, *FEBS Lett.* **69:**108-110.

Luzikov, V. N., L. V. Novikova, A. N. Tikhonov, and A. S. Zubatov, 1983, Correlation between the rate of proteolysis of mitochondrial translation products and fluidity of the mitochondrial inner membrane in Saccharomyces cerevisiae yeast, *Biochem. J.* **214:**785-794.

Luzikov, V. N., and L. V. Romanisha, 1972, Studies on stabilization of the oxidative phosphorylation system. II. Electron transfer-dependent resistance of succinate oxidase and NADH oxidase of submitochondrial particles to proteinases and cobra venom phospholipase, *Biochim. Biophys. Acta* **267:**37-47.

Luzikov, V. N., A. S. Zubatov, E. I. Rainina, and L. E. Bakeyeva, 1971, Degradation and restoration of mitochondria upon deaeration and subsequent aeration of aerobically grown *saccharomyces cerevisiae* cells, *Biochim. Biophys. Acta* **245:**321-334.

Malek, L., L. Bogorad, A. R. Ayers, and A. L. Goldberg, 1984, Newly synthesized proteins are degraded by an ATP-stimulated proteolytic process in isolated pea chloroplasts, *FEBS Lett.* **166:**253-157.

Manjunath, C. K., G. Padmanaban, and H. R. Cama, 1979, Characterization of mitochondrial neutral protease activity and the response of lysosomal enzymes to clofibrate feeding in rat liver, *Biochem. Pharm.* **28:**2929-2934.

Mayer, R. J., 1979, Turnover of mitochondrial proteins in rat liver, *Biochem. Society Trans.* **7:**306-310.

McAda, P. C., and M. G. Douglas, 1981, A neutral metallo endoprotease involved in the processing of an F_1-ATPase subunit precursor in mitochondria, *J. Biol. Chem.* **257**(6):3177-3182.

Mela, L., C. W. Goodwin, and L. D. Miller, 1976, *In vivo* control of mitochondrial enzyme concentrations and activity by oxygen, *Am. J. Physiol.* **231:**1811-1816.

Mela, L., L. D. Miller, L.V. Bacalzo, Jr., K. Olofsson, and R. R. White, 1973, Role of intracellular variations of lysosomal enzyme activity and oxygen tension in mitochondrial impairment in endotoxemia and hemorrhage in the rat, *Ann. Surg.* **178**(6):727-735.

Miura, S., M. Mori, Y. Amaya, and M. Tatibana, 1982, A mitochondrial protease that cleaves the precursor of ornithine carbamoyltransferase, *FEBS Lett.* **106:**641-647.

Mori, M., S. Miura, M. Tatibana, and P. P. Cohen, 1980, Characterization of a protease apparently involved in processing of pre-ornithine transcarbamylase of rat liver, *Natl. Acad. Sci. (USA) Proc.* **77:**7044-7048.

Muller, M., W. Dubiel, J. Rathmann, and S. M. Rapoport, 1980, Determination and characteristics of energy-dependent proteolysis in rabbit reticulocytes, *Eur. J. Biochem.* **109:**405-410.

Nascimento, C. G., G. Jacob, R. Tellez, R. Ocasio, and W. Torres. 1983, Differential effects of ATP on the degradation of mitochondrial translation products synthesized in yeast, *Fed. Proc.* **42:**1813.

Navarro, A., and S. Grisolia, 1984, ATP and other purine nucleotides stimulate the inactivation of ornithine transcarbamylase by broken lysosomes, *FEBS Lett.* **167:**259-262.

Nedergaard, J., and O. Lindberg, 1982, The brown fat cell, *Internat. J. Cytol.* **74:**187-286.

Neri, G., A. M. Gambino, G. Mazzochi, and G. G. Nussdorfer, 1978, Investigation into the effects of ACTH on the half-life of mitochondrial proteins in rat adrenal cortex, *Experientia* **34:**133-134.

Neupert, W., and G. Schatz, 1981, How proteins are transported into mitochondria, *Trends Biochem. Sci.* **6:**1-5.

Nicoletti, M., C. Guerri, and S. Grisolia, 1977, Turnover of Carbamoyl-phosphate Synthase of other mitochondrial enzymes and of rat tissues, *Eur. J. Biochem.* **75:**583-592.

Novikova, L. A., A. S. Zubatov, and V. N. Luzikov, 1981, Multiplicity of mitochondrial proteinases in yeast, *FEBS Lett.* **135:**245-248.

Nussdorfer, G. G., G. Mazzochi, P. Rebuffat, A. S. Bellni, G. Gothardo, and A. M. Gambino, 1975, Investigations on the turnover of adrenocortical mitochondria III, *Cell Tissue Res.* **163:**273-290.

Nussdorfer, G. G., G. Neri, and G. Mazzochi, 1978, Investigations on the turnover of adrenocortical mitochondria XII, *Anat. Rec.* **192:**435-440.

Okada, Y., and T. Omura, 1978, Synthesis and turnover of microsomal and mitochondrial NADH-cytochrome b_5 reductases in rat liver, *J. Biochem.* **83:**1039-1048.

Rajwade, M.S., S. S. Katyare, P. Fatterpaker, and A. Sreenivasan, 1975, Regulation of mitochondrial protein turnover by thyroid hormone(s), *Biochem. J.* **152:**379-387.

Rapoport, S. M., W. Dubiel, and M. Muller, 1981, The mechanism of maturation-dependent breakdown of mitochondria in reticulocytes, *Acta Biol. Med. Germ.* **40:**1277-1283.

Rapoport, S., W. Dubiel, and M. Muller, 1982, Characteristics of an ATP-dependent proteolytic system of rat liver mitochondria, *FEBS Lett.* **147:**93-96.

Rapoport, S., W. Dubiel, and M. Muller, 1983, Calcium exerts an indirect effect on ATP-dependent proteolysis of rat liver mitochondria, *FEBS Lett.* **160:**134-136.

Rapoport, S. M., T. Schewe, R. Wiesner, W. Halangk, P. Ludwig, M. Janicke-Hohne, C. Tannert, C. Hiebsch, and D. Klatt, 1979, The lipoxygenase of reticulocytes, *Eur. J. Biochem.* **96:**545-561.

Raymond, Y., and G. C. Shore, 1980, Processing of the precursor for the mitochondrial enzyme, carbamyl phosphate synthetase, *J. Biol. Chem.* **256:**2087-2090.

Rubio, V., and S. Grisolia, 1977, Prominent role of lysosomes in the proteolysis of rat liver mitochondria at neutral pH, *FEBS Lett.* **75:**281-284.

Russell, S. M., R. J. Burgess, and R. J. Mayer, 1980, Protein degradation in rat liver during post-natal development, *Biochem. J.* **192:**321-330.

Russell, S. M., R. J. Burgess and R. J. Mayer, 1982, Protein degradation in rat liver. Evidence for populations of protein degradation rates in cellular organelles, *Biochim. Biophys. Acta* **714:**34-45.

Russell, S. M., and R. J. Mayer, 1983, Degradation of transplanted rat liver mitochondrial-outer-membrane proteins in hepatoma cells, *Biochem. J.* **216:**163-175.

Saltzgaber-Muller, J., and G. Schatz, 1978, Heme is necessary for the accumulation and assembly of cytochrome c oxidase subunits in Saccharomyces cerevisae, *J. Biol. Chem.* **253:**305-310.

Schatz, G. 1981, The biogenesis of mitochondria, in *Mitochondria and Microsomes,* C. P. Lee, G. Schatz, and G. Dallner, eds., Addison-Wesley, Reading, Mass., pp. 45-66.

Schewe, T., W. Halangk, Ch. Hiebsch, and S. M. Rapoport, 1975, A lipoxygenase in rabbit reticulocytes which attacks phospholipids and intact mitochondria, *FEBS Lett.* **60:**149-152.

Schneider, D. L., 1981, ATP-dependent acidification of intact and disrupted lysosomes. Evidence for an ATP driven proton pump, *J. Biol. Chem.* **256:**3858-3864.

Shaffer, J. B., and M. R. Felder, 1983, Turnover of cytoplasmic and mitochondrial aspartate aminotransferase isozymes in mouse liver and transplantable hepatomas, *Arch. Biochem. Biophys.* **223:**649-661.

Stocco, D. M., J. Cascarano, and M. A. Wilson, 1976, Quantitation of mitochondrial DNA, RNA and protein in starved and starved-refed rat liver, *J. Cell Physiol.* **90:**295-306.

Subramanian, M., S. S. Katyare, P. Fatterpaker, and A. Sreenivasan, 1975, A protease in rat liver mitochondria: Its physiological significance, *Indian J. Biochem. Biophys.* **12:**307-310.

Swamy, KHS., and A. L. Goldberg, 1981, *Eschericia coli* contains eight soluble proteolytic activities, one being ATP-dependent, *Nature* **292:**652-654.

Swick, R. W., A. K. Rexroth, and J. L. Stange, 1968, The metabolism of mitochondrial proteins. III. The dynamic state of rat liver mitochondria, *J. Biol. Chem.* **213:**3581-3587.

Swift, H., and Z. Hruban, 1964, Focal degradation as a biological process, *Fed. Proc.* **23:**1026-1037.

Tanaka, K., L. Waxman, and A. L. Goldberg, 1983, ATP served two distinct roles in protein degradation in reticulocytes, one requiring and one independent of ubiquitin, *J. Cell Biol.* **96:**1580-1585.

Tanaka, K., L. Waxman, and A. L. Goldberg, 1984, Vanadate inhibits the ATP-dependent degradation of proteins in reticulocytes without affecting ubiquitin conjugation, *J. Biol. Chem.* **259:**2803-2813.

Timoneda, J., R. Wallace, and S. Grisolia, 1981, Isolation of a mitochondrial factor from rat liver which potentiates the inactivation of glutamate dehydrogenase by lysosomes, *Biochem. Biophys. Res. Commun.* **101:**555-562.

Umezawa, H., 1976, Structures and activities of protease inhibitors of microbial origin, *Methods Enzymol.* **45:**678-695.

Walker, J. H., R. J. Burgess and R. J. Mayer, 1978, Relative rates of turnover of subunits of mitochondrial proteins, *Biochem. J.* **176:**927-932.

Waxman, L., 1981, Calcium-activated proteases in mammalian tissues, *Methods Enzymol.* **80:**664-680.

Waxman, L., and A. L. Goldberg, 1982, Protease La from *Eschericia coli* hydrolyzes ATP and proteins in a linked fashion, *Natl. Acad. Sci. (USA) Proc.* **79:**4883-4887.

Waxman, L., and A. L. Goldberg, 1985, Protease La, the lon gene product, cleaves specific fluorogenic peptides in an ATP-dependent reaction, *J. Biol. Chem.* **260:**12022-12028.

Waxman, L., J. M. Fagan, K. Tanaka, and A. L. Goldberg, 1985, A soluble ATP dependent system for protein degradation from murine erythroleukemia cells, *J. Biol. Chem.* **260:**11994-12000.

Wheeldon, L. W., and A. L. Lehninger, 1966, Energy-linked synthesis and decay of membrane proteins in isolated rat liver mitochondria, *Biochemistry* **5:**3533-3545.

Wheeldon, L. W., A. Dianoux, M. Bof, and P. V. Vignais, 1974, Stable and labile products of mitochondrial protein synthesis in vitro, *Eur. J. Biochem.* **46:**189-199.

Woodbury, R. G., G. M. Gruzenski, and D. Lagunoff, 1978, Immunofluorescent localization of a serine protease in rat small intestine, *Natl. Acad. Sci. (USA) Proc.* **75:**2785-2789.

Woodrow, G., and G. Schatz, 1979, The role of oxygen in the biosynthesis of cytochrome c oxidase of yeast mitochondria, *J. Biol. Chem.* **254:**6088-6093.

Zak, R., A. F. Martin, and R. Blough, 1979, Assessment of protein turnover by use of radioisotopic tracers, *Physiol. Rev.* **59:**407-446.

Zehnbauer, B. A., E. C. Foley, G. W. Henderson, and A. Markovitz, 1981, Identification and purification of the lon (cap r) gene product, a DNA-binding protein, *Natl. Acad. Sci. (USA) Proc.* **78:**2043-2047.

3

Perinatal Development of Mitochondria in Rat Liver

June R. Aprille

Tufts University

INTRODUCTION AND HISTORICAL PERSPECTIVE

In the fetus the liver is primarily an hematopoietic organ whereas in neonatal life it has a plethora of metabolic functions. It is naturally interesting to ask how and when during development this transition occurs. Of course, there is an actual change in cell type during late gestation that must be considered, but in spite of the predominance of parenchymal cell mass by the time of birth (Herzfeld, Federman, and Greengard, 1973), specialized metabolic activities such as fatty acid oxidation, gluconeogenesis, and ureogenesis are not yet developed. Mitochondria play an important part in the development of these pathways, not only because of their essential bioenergetic role but also because certain key enzymes of each pathway are localized in the matrix compartment. The purpose of this review is to provide an integrated overview of perinatal changes in mitochondrial morphology, enzyme composition, and function that contribute to metabolic development in the liver.

Mitochondria were not unequivocally identified as the subcellular location of respiratory enzymes until the late 1940s and early 1950s. Therefore, it is not surprising that there is no reference to mitochondria in the first developmental study of respiratory enzymes in brain and liver homogenates, reported in 1945 by Potter, Schneider, and Liebl. These authors were interested in potential similarities of bioenergetic function between undifferentiated normal tissue and tumors, a theme that remains current. Since then the development of individual enzyme activities in mitochondria has been described in many reports and from many points of view. Pollak and Duck-Chong surveyed the available literature on rat liver in 1973. Some aspects of mitochondrial function have been included in more general reviews on mechanisms of enzyme development or on the development of metabolic competence (Pollak and Sutton, 1980*a;* Herrmann and Tootle, 1964; Dawkins, 1966; Greengard, 1971; Böhme, Sparmann, and Hofmann, 1983; Hahn and Skala, 1971; Bittner et al., 1979; Jones, 1982), but there is no recent review that focuses exclusively on mitochondrial development in mammalian liver.

Lately there has been increasing appreciation of the mitochondrion as a closed, selectively permeable compartment. Investigation into transport phenomena and the matrix concentrations of metabolic intermediates and cofactors was recognized by Hommes in 1975 as necessary for a complete understanding of mitochondrial function during development. In this review particular attention will be given to recent findings showing that substrate access to the matrix is an important regulator of mitochondrial enzyme activities in the perinatal period.

PERINATAL DEVELOPMENT AS A MODEL FOR MITOCHONDRIAL BIOGENESIS

Mitochondrial development in the liver involves changes in the number of mitochondria per cell and changes in the size of the mitochondrial compartment relative to cell volume. Besides the obvious significance that these phenomena have for the development of liver metabolism, they are also interesting as a model of mitochondrial biogenesis in a mammalian system (Pollak and Duck-Chong, 1973; Pollak and Sutton, 1980*a*). During perinatal development, biogenesis does result in proliferation, but changes in the size, ultrastructure, and enzyme composition of mitochondria occur as well. Throughout the discussion, observations relevant to both aspects of mitochondrial development (proliferation and differentiation) will be pointed out, and therefore some general comments and definitions will be helpful as a background.

Mitochondrial Proliferation

Proliferation is a term that should be restricted to mean an increase in number. Therefore, strictly speaking, proliferation of mitochondria can be assessed only by counting mitochondria in electron micrographs since increases in chemical or enzyme markers (DNA, cardiolipin, cytochrome oxidase, etc.) could be attributed to a relative increase in organelle size as well as in number. Results can be expressed in various ways (e.g., number per cell or number per unit cytoplasm) but the data should not be interpreted too rigidly because in living cells mitochondria undergo fusion and fission and move within the cytoplasm (Wildman, Hongladarom, and Honda, 1962; Chen et al., 1982). Nevertheless, upon fixation of any given tissue, mitochondria do assume reproducible characteristics of size, shape, and relative abundance that are measurable. Quantitative stereometric methods can also be used to estimate the size of the total mitochondrial compartment relative to cell volume (usually called the volume fraction or volume density) (Herzfeld, Federman, and Greengard, 1973; Rohr et al., 1971; David, 1979; Lang and Herbener, 1972).

The morphometric characteristics of mitochondrial development in rat liver are well established. An active period of proliferation (at a rate faster than cell growth) occurs immediately after birth. In general, proliferation probably is viewed best as a sequence of events that includes molecular synthesis, assembly, and organelle division occurring over a period of time in a manner somewhat analogous to cell replication. Certainly the many separate lipid, protein, and nucleic acid components of mitochondria are not made and assembled simultaneously. Chronological correlation of the morphological and biochemical changes that occur during the postnatal phase of proliferation might reveal the actual order of events and suggest regulatory mechanisms for mitochondrial biogenesis.

Mitochondrial Differentiation

In theory, the simplest outcome of proliferation is an expanded population of mitochondria identical to the parent population. During development a more typical result is that the new mitochondria are modified with respect to ultrastructure, composition, enzyme specific activities, and function. Changes in these characteristics can be defined as mitochondrial differentiation. Differentiation can also be envisioned to occur in the absence of any proliferation by alterations of the existing mitochondria.

The concept of organelle differentiation can apply to bioenergetic function in that mitochondria from various tissues may have varying respiratory rates. The idea is most readily applied, however, to developmental changes in ultrastructure and to the development of special metabolic functions. Mitochondria usually acquire enzymes specialized for the cell type in which they happen to be found. For example, liver mitochondria are outfitted with specific enzymes that endow the ability to oxidize fatty acids, generate ketone bodies, and carry out reactions that are part of major biochemical pathways such as gluconeogenesis and urea synthesis. These special enzyme complements are invariably nuclear gene products superimposed onto the basic organelle design. They often develop on a schedule relatively independent of the events associated with mitochondrial proliferation.

One way to assess the relative contributions of organelle proliferation and differentiation is to compare the biochemical measurement of interest in a homogenate and in a mitochondrial fraction. For example, any developmental increase in enzyme activity observed in homogenates might be attributed to proliferation if the activity per milligram protein remains constant in isolated mitochondria as a function of developmental age. On the other hand, the extent to which specific activity of the enzyme in isolated mitochondria does change probably reflects altered characteristics of the organelle, that is, mitochondrial differentiation. If this approach is followed, it is important to be sure that the relative purity of the isolated mitochondrial fraction is equivalent at each developmental age studied. Also, proliferation must be verified by morphometric methods because an increase in specific activity in homogenates might be due to increased organelle size rather than number. Finally, it should be noted again that these distinctions are important only with respect to understanding biogenesis. From a metabolic point of view, an increase in mitochondrial size may be functionally equivalent to an increase in number.

NORMALIZATION OF BIOCHEMICAL ACTIVITIES OF THE LIVER MEASURED DURING DEVELOPMENT

During fetal life, 7 to 10% of liver wet weight is protein (Van Rossum, 1963; Krahling et al., 1979; Mertens-Strijthagen et al., 1979; Lang, 1965); this value increases to 17% at the time of birth due to postnatal glycogen depletion (Böhme, Sparmann, and Hofmann, 1983; Rohr et al., 1971; Shelley and Neligan, 1966; Greengard, Federman, and Knox, 1972; Villee and Hagerman, 1958) and to a 20% reduction in extracellular fluid volume (Rohr et al., 1971; Vergonet, Hommes, and Molenaar, 1970). If one is interested in the perinatal development of particular enzymes relative to other cell proteins, it thus is inappropriate to express activity data relative to wet or dry weight of the tissue. Cellular DNA is also a poor normalizing factor for this purpose for two reasons—there is a postnatal wave of cell division for which equivalent cell growth is not attained for at least two weeks, and liver cells have variable ploidy. Of the various alternatives (Herzfeld, Federman, and Greengard, 1973; Pollak and Duck-Chong, 1973), expressing activities relative to tissue protein is probably the best means of discerning specific increases in mitochondrial enzymes over other cellular proteins in liver. Therefore, in comparing various reports for this review, values were recalculated as necessary.

Differences in the relative purity of mitochondria isolated at successive ages is a potential problem that plagues developmental studies in some organs. Isolated liver mitochondria from late fetal and newly born (< 1 hour) animals do pellet more loosely and band at greater densities on sucrose gradients than at later postnatal ages (Krahling et al., 1979; Mertens-Strijthagen et al., 1979; Pollak and Munn, 1970; Hallman and Kankare, 1971). Fortunately, this behavior is very predictable. For liver it has been shown that the relative contamination of mitochondrial fractions by lysosomes and peroxisomes does not vary much as a function of age when a standard method of isolation by differential or gradient centrifugation is used (Mertens-Strijthagen et al., 1979). If necessary, purity can be monitored with a stable marker such as cardiolipin (Pollak and Munn, 1970; Hallman and Kankare, 1971, 1979; Hostetler, Zenner, and Morris, 1979; Jakovcic et al., 1971) or by electron microscopy.

DEVELOPMENTAL CHANGES IN LIVER CELL POPULATIONS AND MITOCHONDRIAL MORPHOLOGY

As noted before, the biochemical development of mitochondria in the liver is likely to be more interesting if viewed against the background of morphological development. Most of the observations described in this section were made by quantitative stereometric analysis of electron micrographs. A variety of fixatives were used (osmium, glutaraldehyde, and phosphotungstic acid) with consistent findings among the several reports, so that it was possible to construct the composite diagram in Figure 3-1 with confidence.

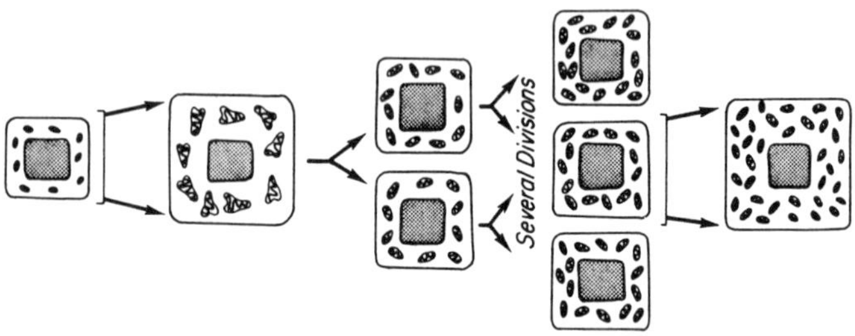

AGE:	18 Day Fetus →	Term Fetus →	1-2 Days →	12 Days →	28 Days
MITOCHONDRIA:					
Number per Cell:	300	300	350	500	700-1500
Volume Fraction:	10%	13%	21%	21%	21%
Ave. Volume (μ^3):	1.0	2.5	1.8	1.8	1.2
HEPATOCYTE VOL.:		↑3X	↑50%	No Change	↑2.5X
ORGAN WEIGHT:		↑3X	↑sl.	↑3X	↑2.5X

Figure 3-1. Morphometry of hepatocytes and mitochondria in developing rat liver.

Figure 3-1 indicates perinatal changes in the number of mitochondria per cell (Rohr et al., 1971; David, 1979; Lang and Herbener, 1972; Vergonet, Hommes, and Molenaar, 1970; Gregson and Williams, 1969); mitochondrial volume fraction, determined stereometrically from electron micrographs as the percentage of points overlaying the cytoplasm that fall on mitochondria (Herzfeld, Federman, and Greengard, 1973; David, 1979; Lang and Herbener, 1972; Vergonet, Hommes, and Molenaar, 1970); mitochondrial volume, μ^3, calculated from average length and width assuming a sphere or cylinder (Herzfeld, Federman, and Greengard, 1973; Rohr et al., 1971; David, 1979; Lang and Herbener, 1972; Jakovcic et al., 1971; Gregson and Williams, 1969); relative hepatocyte volume (Rohr et al., 1971; Greengard, Federman, and Knox, 1972; Vergonet, Hommes, and Molenaar, 1970); and total organ weight (Rohr et al., 1971; Greengard, Federman, and Knox, 1972) as a function of age. Values are composite estimates from the references cited. Note that between 18 days gestation and term, cell volume triples. The number of mitochondria per cell remains constant, but individual mitochondria enlarge so that the volume fraction does not change. After birth, mitochondria are reduced in size by fission, and further proliferation results in an increase in the volume fraction within the first day or two. Then mitochondrial proliferation continues, but at a rate that just keeps pace with cell division and growth. Thus the number of mitochondria per cell increases but the volume fraction remains constant as development proceeds. These events are discussed below in more detail.

Prenatal Period

During most of gestation, hematopoietic cells are the major cell type in the liver. Parenchymal cells are much larger, however, so that at 18 days gestation the cells already account for about 60% of the cytoplasmic volume in the organ (Greengard, Federman, and Knox, 1972). At this point, parenchymal cell size is only one third of adult size and the nucleus occupies 41% of the cell volume (Fig. 3-1) (Vergonet, Hommes, and Molenaar, 1970). Mitochondria have about the same average dimensions (Herzfeld, Federman, and Greengard, 1973; Rohr et al., 1971) as they do in the weanling or adult (Herzfeld, Federman, and Greengard, 1973; David, 1979; Lang and Herbener, 1972; Jakovcic et al., 1971; Gregson and Williams, 1969). The fraction of cytoplasmic volume occupied by mitochondria is still only about one half (Herzfeld, Federman, and Greengard, 1973; Rohr et al., 1971; Vergonet, Hommes, and Molenaar, 1970) of that in adults (Herzfeld, Federman, and Greengard, 1973; David, 1979; Lang and Herbener, 1972; Vergonet, Hommes, and Molenaar, 1970) because the number of mitochondria per unit of cytoplasm is much lower (Fig. 3-1) (Rohr et al., 1971).

Between 18 days gestation and term, parenchymal cell size triples (Fig. 3-1) so that by the time of birth these cells account for 85% of the total cytoplasmic area of liver (Greengard, Federman, and Knox, 1972) and the weight of the whole organ has increased 2-fold (Rohr et al., 1971). During this interval the linear dimensions of mitochondria increase 1.4-fold, which more than doubles the individual volume (Rohr et al., 1971; David, 1979; Lang and Herbener, 1972; Jakovcic et al., 1971; Gregson and Williams, 1969). Shapes become more varied (David, 1979; Lang and Herbener, 1972) but the number of mitochondria per unit cytoplasm remains constant (Rohr et al., 1971). The proliferation plus enlargement of mitochondria apparently combine to allow an increase in the total mitochondrial compartment that barely keeps pace with cell growth because the volume fraction of mitochondria increases very little (Fig. 3-1) (Herzfeld, Federman, and Greengard, 1973; Rohr et al., 1971). The density of cristae per unit cell area increases 1.8-fold between 18 days gestation and term (Rohr et al., 1971), suggesting that the amount of inner membrane within the mitochondrial compartment increases in proportion to the increase in cross-sectional area of the organelle.

Postnatal Period

Within one day after birth, parenchymal cells begin to divide and the average cell size is reduced by half (Fig. 3-1) (Rohr et al., 1971; Greengard, Federman, and Knox, 1972). Division continues with no change in cell size until day 12; by that time organ weight has increased threefold (Rohr et al., 1971; Greengard, Federman, and Knox, 1972). Between days 12 and 28, cellular hypertrophy dominates the picture, resulting in a twofold to threefold increase both in cell size and in organ weight (Greengard, Federman, and Knox, 1972). A similar pattern of postnatal hyperplasia followed by hypertrophy occurs in rabbit liver (Milner and Pownall, 1981).

Very soon after birth, perhaps within minutes or hours, the average size (volume) of individual mitochondria decreases by about 30%, probably by fission of the large mitochondria existing at term (Fig. 3-1) (Rohr et al., 1971; Lang and Herbener, 1972). Also, with the onset of postnatal cell division, these smaller mitochondria proliferate rapidly, contributing to a twofold increase both in the volume fraction (Herzfeld, Federman, and Greengard, 1973; Rohr et al., 1971; David, 1979) and in the number of mitochondria per unit cytoplasm within a day or two (Rohr et al., 1971). After that, the number of mitochondria per cell continues to increase, but only in proportion to cell growth and division (Rohr et al., 1971; David, 1979; Lang and Herbener, 1972; Vergonet, Hommes, and Molenaar, 1970; Gregson and Williams, 1969). Thus, the mitochondrial volume fraction remains constant after a few days (Herzfeld, Federman, and Greengard, 1973; Rohr et al., 1971; David, 1979) even though the number of mitochondria per cell increases steadily until mature cell size is attained at four to five weeks of age. The relative abundance of cristae within each mitochondria does not change postnatally (Rohr et al., 1971; Jakovcic et al., 1971) but the electron density of the matrix increases (Mintz et al., 1967), as does the amount of protein per mitochondrion (Herzfeld, Federman, and Greengard, 1973).

ENZYMES OF OXIDATIVE METABOLISM

In most studies and reviews the development of Krebs Cycle and electron transport enzymes in rat liver homogenates is usually treated as a continuous phenomenon extending from late gestation through postnatal life. It is possible, however, that some enzymes develop on a biphasic schedule related to the distinctly different morphological changes that characterize mitochondrial development before and after birth, and so it is important to consider the prenatal and postnatal periods separately.

Enzyme Development During Late Fetal Life

Relatively few studies have included systematic measurements of mitochondrial enzyme activities in the preterm animal beginning as early as 18 days gestation. The activities of the following enzymes have been measured occasionally in prenatal liver, but they need to be studied further to establish a clear picture of *in utero* development: pyruvate dehydrogenase (Knowles and Ballard, 1974; Berger and Hommes, 1975); aconitase (Levy and Toury, 1970; Hommes, Luit-de Hann, and Richters, 1971); NAD-dependent malate dehydrogenase (Mertens-Strijthagen et al., 1979; Levy and Toury, 1970; Hallman, 1971*a;* Dawkins, 1959); citrate synthetase (Hommes, Luit-de Hann, and Richters, 1971); α-ketoglutarate dehydrogenase (Hommes, Luit-de Hann, and Richters, 1971); fumarase (Levy and Toury, 1970; Hommes, Luit-de Hann, and Richters, 1971); NAD- and NADP-dependent isocitrate dehydrogenase (Levy and Toury, 1970; Dawkins, 1959); NADH dehydrogenase (Levy and Toury, 1970); succinic and NADH oxidases (Levy and Toury, 1970); NADH- and succinate-cytochrome c reductases (Lang, 1965; Jakovcic et al., 1971; Dawkins, 1959); and the chemical amounts of cytochromes aa_3, b, cc_1, and flavoprotein (Hallman, 1971; DeVos, Wilmink, and Hommes, 1968). The study of some of these enzymes is complicated by the existence of cytosolic and mitochondrial isozymes (Hommes, Luit-de Hann, and Richters, 1971; Dawkin; 1959; Vernon and Walker, 1968).

For the last three to four days of gestation, there is reasonably consistent developmental data from several laboratories for only three mitochondrial enzymes: cytochrome oxidase (Lang, 1965; Jakovcic et al., 1971), succinate dehydrogenase (Potter, Schneider, and Liebl, 1945; Jakovcic et al., 1971; Hallman, 1971*a*), and NAD-dependent glutamate dehydrogenase (Herzfeld, Federman, and Greengard, 1973; Hommes and Richters, 1969). Figure 3-2 shows the prenatal pattern of development for these three enzymes in whole homogenates (panel *A*) and in isolated mitochondria (panel *B*) as follows: cytochrome oxidase (———), succinate dehydrogenase (— —), and NAD-dependent glutamate dehydrogenase (· · · · ·). The development profiles are composites derived from values reported in the references just cited. To emphasize the relative magnitude of changes, specific activities are plotted relative to their values at term, set arbitrarily at 1.0.

From day 18 to birth, mitochondria increase in size but not in number per cell or in volume fraction (Fig. 3-1). The amount of mitochondrial protein isolated per gram liver also remains constant (Herzfeld, Federman, and Greengard, 1973), yet during this interval, the specific activity of cytochrome oxidase decreases by more than 50% and then rebounds just before birth in both homogenates and isolated mitochondria (Fig. 3-2*A* and 3-2*B*) (Herzfeld, Federman, and Greengard, 1973; Jakovcic et al., 1971; Lang, 1965). Succinate dehydrogenase (Potter, Schneider, and Liebl, 1945; Jakovcic et al., 1971) activity increases slightly, and NAD-dependent glutamate dehydrogenase activity approximately doubles (Fig. 3-2*A* and 3-2*B*) (Herzfeld, Federman, and Greengard, 1973; Knowles and Ballard, 1974; Hommes and Richters, 1969; but see also Hallman, 1971*a*). The enzyme changes are such that they cannot be readily accounted for by the increasing proportion of parenchymal cells in the tissue (Herzfeld, Federman, and Greengard, 1973). These findings suggest that while mitochondrial mass is increasing at a rate

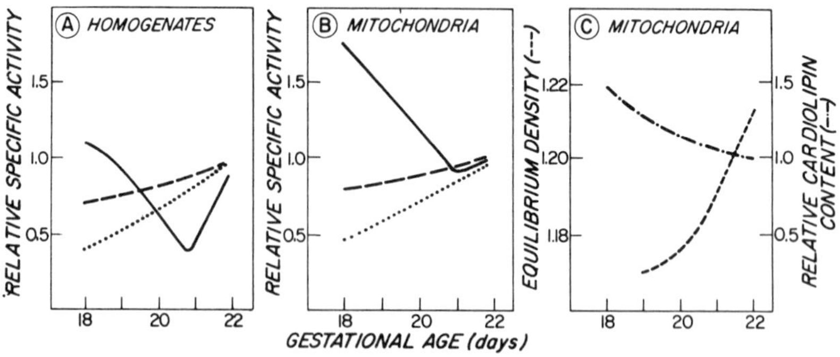

Figure 3-2. Mitochondrial development during late gestation. Panels *A* (homogenates) and *B* (mitochondria) show cytochrome oxidase (———), succinate dehydrogenase (- - - -), and NAD-dependent glutamate dehydrogenase (· · · · ·). Panel *C* shows equilibrium density (— — — —) and cardiolipin content (— · — ·) of isolated mitochondria. See text for explanation.

commensurate with parenchymal cell growth (Fig. 3-1), the morphology and enzyme composition of the organelle is changing.

In Figure 3-2, panel *C* shows cardiolipin content (— · — ·) (Jakovcic et al., 1971) and equilibrium density (major peak on an isopycnic sucrose gradient) (— — —) (Pollak and Munn, 1970; Hallman and Kankare, 1971) for isolated mitochondria in relation to gestational age, as a composite of data reported in the references just cited. Cardiolipin content is shown relative to the value at term, set arbitrarily at 1.0. For equilibrium density, actual measured values are shown. The equilibrium density of mitochondria steadily increases during late gestation (Pollak and Munn, 1970; Hallman and Kankare, 1971), another indication of changing organelle composition (Fig. 3-2*C*). The phospholipid composition of mitochondrial membranes does not change much at any time during development (Hallman and Kankare, 1971, 1979; Hostetler, Zenner, and Morris, 1979; Jakovcic et al., 1971; Miyahara et al., 1984; Meister et al., 1983). It has been suggested that the greater density is due to greater protein concentration in the whole organelle relative to membrane content (Mertens-Strijthagen et al., 1979; Hallman and Kankare, 1971). Cardiolipin per mg protein remains constant in homogenates but decreases by about 40% in isolated mitochondria (Fig. 3-2*C*) (Jakovcic et al., 1971) consistent with the idea that the organelle develops a lower lipid/protein ratio during the last four to five days of gestation. This may simply be a consequence of the lower ratio of inner membrane area to matrix volume that accompanies the increase in organelle size during late gestation. The relative abundance of cristae increases 1.8-fold (Rohr et al., 1971), whereas the organelle volume increases 2.5-fold (Fig. 3-1), a predictable result based on the different geometrical formulas for the area (πr^2) and volume ($4/3\pi r^3$). The greater increase in glutamate dehydrogenase (a matrix enzyme) relative to cytochrome oxidase and succinate dehydrogenase (membrane bound enzymes) may also be a consequence of this geometric relationship, although other interpretations are just as likely. For example, glutamate dehydrogenase activity is a source of ammonia (Krebs et al., 1978) and so its development may be coordinated with the development of the urea cycle. Furthermore, succinate dehydrogenase and cytochrome oxidase have differing developmental patterns (Fig. 3-2) even though both are membrane enzymes.

Overall, the evidence suggests that late gestation is characterized by biochemical and morphological differentiation of the organelle. As the mitochondria enlarge, there appears to be an increase in matrix volume relative to inner membrane surface area. The limited data available give the impression that some oxidative enzymes may be relatively enriched and some are decreased within the structural unit. The significance of this differentiated state may be that the organelle population is prepared for the rapid proliferation that follows soon after birth.

Enzyme Development During Postnatal Life

Within a day or two of birth, the volume fraction of mitochondria within the cytoplasm increases by 1.5 to 2-fold, while average mitochondrial size decreases (Fig. 3-1). Equilibrium density decreases (Fig. 3-3*C*) (Krahling et al., 1979; Mertens-Strijthagen et al., 1979), possibly because of the increased membrane area relative to matrix space that is a consequence of the smaller average size. During this proliferative phase the total mitochondrial mass appears to increase at a much faster rate than general cellular growth. Consistent with this idea are reports that the amount of mitochondrial protein and markers such as cytochrome oxidase and cardiolipin isolated per gram tissue increases 2- to 3-fold (Herzfeld, Federman, and Greengard, 1973; Mintz et al., 1967; Mertens-Strijthagen et al., 1979; Jakovcic et al., 1971; Gregson and Williams, 1969; Hommes, Luit-de Hann, and Richters, 1971; Hallman, 1971*a;* Dawkins, 1959; Ontko, 1966; Kinnula and Hassinen, 1977).

Also, the homogenate-specific activity of virtually every oxidative enzyme studied so far increases at least 2-fold during the first week of postnatal life (Fig. 3-3*A*). This is true for inner membrane components such as cytochrome oxidase (Herzfeld, Federman, and Greengard, 1973; Potter, Schneider, and Liebl, 1945; Lang and Herbener, 1972; Mertens-Strijthagen et al., 1979; Lang, 1965; Dawkins, 1959; Kinnula and Hassinen, 1977), succinate dehydrogenase (Potter, Schneider, and Liebl, 1945; Hostetler, Zenner, and Morris, 1979; Jakovcic et al., 1971; but see also Gregson and Williams, 1969), and respiratory cytochromes (Kinnula and Hassinen, 1977; Hallman, 1971*b;* Dallman and Schwartz, 1964; Mäenpää and Räihä, 1968), and also for

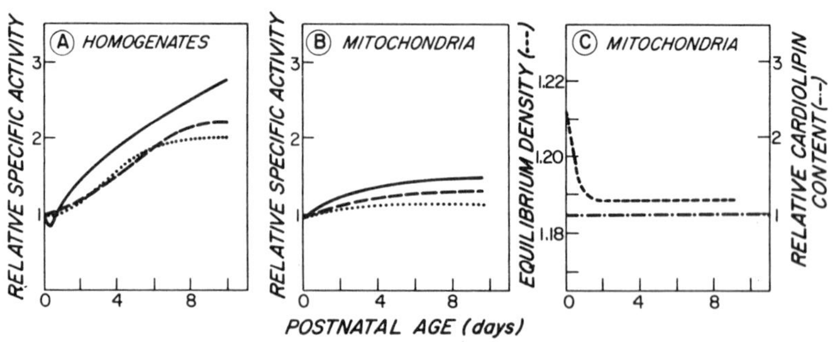

Figure 3-3. Mitochondrial development during postnatal development. Panels *A* (homogenates) and *B* (mitochondria) show cytochrome oxidase (————), succinate dehydrogenase (- - - -), and NAD-dependent glutamate dehydrogenase (· · · · ·). Panel *C* shows equilibrium density (— — — —) and cardiolipin content (— · — ·) of isolated mitochondria. See text for explanation.

NAD-dependent glutamate dehydrogenase (Herzfeld, Federman, and Greengard, 1973; Krahling et al., 1979; Mertens-Strijthagen et al., 1979; Dawkins, 1959; Hommes and Richters, 1969), and total pyruvate dehydrogenase activity (Knowles and Ballard, 1974). For all the enzymes that develop after birth, the fastest rate of change is within the first several days, coincident with the most active period of cell division and mitochondrial proliferation (compare Fig. 3-1 to 3-3*A*).

The data on postnatal enzyme development in isolated mitochondria are not as uniform as for homogenates. In general, though, whenever an enzyme increase is noted in isolated mitochondria, it is of much smaller magnitude than is observed in homogenates (Fig. 3-3*B*). Most authors report postnatal increases of less than 50% in the mitochondrial-specific activities of succinate dehydrogenase (Hostetler, Zenner, and Morris, 1979; Jakovcic et al., 1971; Levy and Toury, 1970; Hallman, 1971*a*), cytochrome oxidase (Mertens-Strijthagen et al., 1979; but see also Jakovcic et al., 1971), cytochrome content (DeVos, Wilmink, and Hommes, 1968; Kinnula and Hassinen, 1977; but see also Hallman and Kankare, 1971 and Hallman, 1971*a*), NAD-glutamate dehydrogenase (Mertens-Strijthagen et al., 1979; Knowles and Ballard, 1974; Levy and Toury, 1970; Hallman, 1971*a*), and a number of Krebs Cycle enzymes (Fig. 3-3*B*) (Levy and Toury, 1970; Hommes, Luit de Hann, and Richters, 1971; Hallman, 1971*a;* DeVos, Wilmink, and Hommes, 1968).

Figure 3-3 summarizes the postnatal pattern of development in whole homogenates *(A)* and in isolated mitochondria *(B)* for three mitochondrial enzymes: cytochrome oxidase (—) (Herzfeld, Federman, and Greengard, 1973; Potter, Schneider, and Liebl, 1945; Mertens-Strijthagen et al., 1979; Lang, 1965; Jakovcic et al., 1971), succinate dehydrogenase (— —) (Potter, Schneider, and Liebl, 1945; Jakovcic et al., 1971; Hallman, 1971*a*), and NAD-dependent glutamate dehydrogenase (· · · ·) (Herzfeld, Federman, and Greengard, 1973; Krahling et al., 1979; Mertens-Strijthagen et al., 1979; Hallman, 1971*a;* DeVos, Wilmink, and Hommes, 1968; Hommes and Richters, 1969). Panel *C* shows cardiolipin content (— · — ·) (Hallman and Kankare, 1971; Jakovcic et al., 1971) and equilibrium density (— — —) (major peak on an isopycnic gradient) (Krahling et al., 1979; Mertens-Strijthagen et al., 1979; Pollak and Munn, 1970; Hallman and Kankare, 1971) for isolated mitochondria as a function of postnatal age. The developmental patterns are composites derived from activities reported in the references cited. To emphasize the relative magnitude of changes, specific activities are plotted relative to their values at term set arbitrarily at 1.0 for the enzymes and cardiolipin. Measured values are displayed for equilibrium density.

Finding large increases in enzyme-specific activities in homogenates but not in isolated mitochondria shows that, at least for the oxidative enzymes, the postnatal increase is due mainly to proliferation rather than to organelle differentiation. This fact was recognized in several early studies (Herzfeld, Federman, and Greengard, 1973; Dawkins, 1966; Jakovcic et al., 1971; Dawkins, 1959) and the proliferative phase of mitochondrial development has been studied as a model for biogenesis. Some evidence suggests that various components of the organelle are assembled sequentially, not simultaneously. For example, the synthesis of basic membrane structure may precede the insertion of active enzymes (Hallman and Kankare, 1971, 1979; Meister et al., 1983). The rate of cardiolipin synthesis is fastest immediately after birth (Hallman and Kankare, 1979); in isolated mitochondria the amount of cardiolipin per mg protein remains constant (Hallman and Kankare, 1971; Jakovcic et al., 1971; Miyahara et al., 1984) while cytochrome aa_3 increases over a period of time (Hallman and Kankare, 1971; Kinnula and Hassinen, 1977). The mitochondrial volume fraction increases to a maximum within a day or two after birth (Fig. 3-1), while the major proliferative change in enzyme activities and protein occurs over a longer period (Herzfeld, Federman, and Greengard, 1973) (Fig. 3-3*A*); also the mitochondrial specific activities of many enzymes continue to increase slowly for several days (Fig. 3-3*B*). Finally, the activities of citrate synthetase and cytochrome oxidase in isolated mitochondria appear to be transiently decreased for several hours following birth (Escriva et al., 1984), possibly indicating that the mitochondrial enzymes are diluted by rapid membrane bilayer assembly during the most active phase of structural proliferation.

Summary of Oxidative Enzyme Development

During the last four days of gestation, individual mitochondria enlarge in size and the ratio of matrix volume to inner membrane increases. The average density of the organelle increases while cardiolipin per mg mitochondrial protein decreases. The volume fraction and cardiolipin per mg tissue protein remain constant. All of these changes are consistent with a model of organelle differentiation in that the mitochondria are altered with respect to structure and enzyme composition. There is no obvious metabolic significance of the differentiated state, but the changes seem to result in large organelles poised for a rapid proliferative response.

Immediately after birth, possibly within a few hours, the average density of mitochondria decreases. Cardiolipin synthesis is at a maximum, and the volume fraction of mitochondria doubles. This appears to represent expansion of the mitochondrial compartment both by fission of existing organelles and by rapid assembly of new membrane. Oxidative enzyme activities in isolated mitochondria may be transiently diluted, but then are quickly replenished. After a few days, a twofold to threefold proliferative increase in mitochondrial respiratory capacity relative to cell volume is accomplished.

Superimposed over these prenatal and postnatal phases of mitochondrial development is a pronounced increase in several special function enzymes, discussed in the next section.

SPECIAL FUNCTION ENZYMES

Other studies have given general overviews of liver metabolic function in developing mammals (Jones, 1982; Böhme, Sparmann, and Hofmann, 1983; Girard et al., 1981; Bailey, 1981). This discussion will be concerned only with the development of selected key enzymes in the pathways of gluconeogenesis, fatty acid oxidation, and ureogenesis that happen to be localized in mitochondria. For gluconeogenesis, one such enzyme is pyruvate carboxylase (Fig. 3-4*A*) (Snell, 1974; Ballard and Hanson, 1967; Chang, 1977), which also serves an important anaplerotic role for Krebs Cycle intermediates (Bieber et al., 1980). Pyruvate carboxylase is synthesized slowly during late gestation and rapidly in early postnatal life, attaining a maximum concentration in isolated mitochondria by one to two days of age (Fig. 3-4*B*) (Snell, 1974; Ballard and Hanson, 1967; Chang, 1977). The postnatal development of this protein in mitochondria is amplified in the whole cell by the extended period of mitochondrial proliferation characterized earlier (Fig. 3-4*B*). Although substantial amounts of the enzyme are present at term it is not active in intact mitochondria until an hour or so after birth (Snell, 1974) because the enzyme is subject to regulation by the matrix ATP concentration (Aprille, Yaswen, and Rulfs, 1981; Brennan and Aprille, 1984).

In Figure 3-4 panel *A* shows the subcellular compartmentation of enzymes

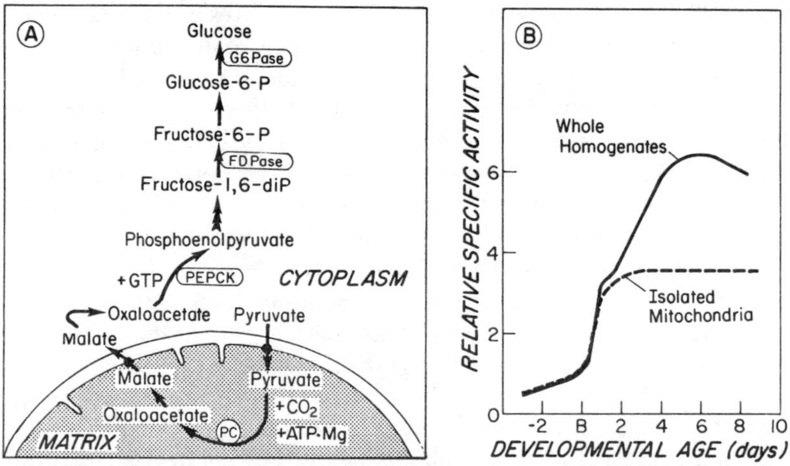

Figure 3-4. Development of pyruvate carboxylase (PC) (panel *B*), a mitochondrial enzyme in the gluconeogenic pathway (panel *A*).

in the gluconeogenic pathway in rat hepatocytes. Pyruvate from the cytoplasm is transported into the mitochondria and converted to oxaloacetate by pyruvate carboxylase (PC) in an ATP-dependent reaction. Oxaloacetate is transported out via the malate shuttle where it continues through gluconeogenesis. The other enzymes of the gluconeogenic pathway, phosphoenolpyruvate carboxykinase (PEPCK), fructose-1,6-diphosphatase (FDPase), and glucose-6-phosphatase (G6Pase), are located in the cytoplasm in the rat, although PEPCK has a mitochondrial isozyme. Panel *B* shows the specific activity of total pyruvate carboxylase in disrupted rat liver mitochondria (— —) and in whole homogenates (———) during development, relative to values of 1.0 at term. The developmental patterns for pyruvate carboxylase are composites derived from activities reported by Snell (1974), Ballard and Hanson (1967), and Chang (1977).

Cytosolic PEPCK activity develops 20- to 30-fold postnatally in the rat (Ballard and Hanson, 1967) and is the rate-limiting enzyme for maximum gluconeogenesis after one to two hours, when pyruvate carboxylase has become active (Ballard, 1971; Aprille, Yaswen, and Rulfs, 1981; Brennan and Aprille, 1984, 1985). Mitochondrial PEPCK has a significant gluconeogenic role in some species (Bieber et al., 1980; Parameswaran and Arinze, 1981), but probably not in the rat or rabbit, where its activity is low and relatively constant throughout development (Ballard and Hanson, 1967; El Manoubi et al., 1983).

The ability to oxidize long chain fatty acids emerges postnatally and is limited by the development of carnitine acyltransferase and to some extent by acyl-CoA synthase (for overview see Bailey, 1981; Chalk et al., 1983). Long chain fatty acids (or their CoA esters) cannot permeate the inner membrane. They are activated to acylCoA derivatives by a synthetase in the outer mitochondrial membrane or in the microsomes, converted to the carnitine ester by transferase A, exchanged across the membrane for free carnitine, then reconverted to the CoA ester by transferase B on the matrix side where they finally undergo beta oxidation (Fig. 3-5*A*). Carnitine acyltransferase does not actually catalyze the transport, but it participates on both the inner and outer faces of the membrane to provide appropriate substrates for the carrier, carnitine-acylcarnitine translocase.

The rapid development of carnitine acyltransferase has adaptive significance for the utilization of a milk diet by the suckling animal. The chain length specificity of carnitine acyltransferase activity in the rat and rabbit appears to be related to differences in the predominant chain length of fatty acids in the milk of each species (Chalk et al., 1983; Rulfs and Aprille, 1980; Rulfs, 1981). Enzyme activity develops to a maximum soon after birth (Fig. 3-5*B*) and decreases after weaning (Chalk et al., 1983). Besides being an important pathway for energy metabolism in the neonate, mitochondrial

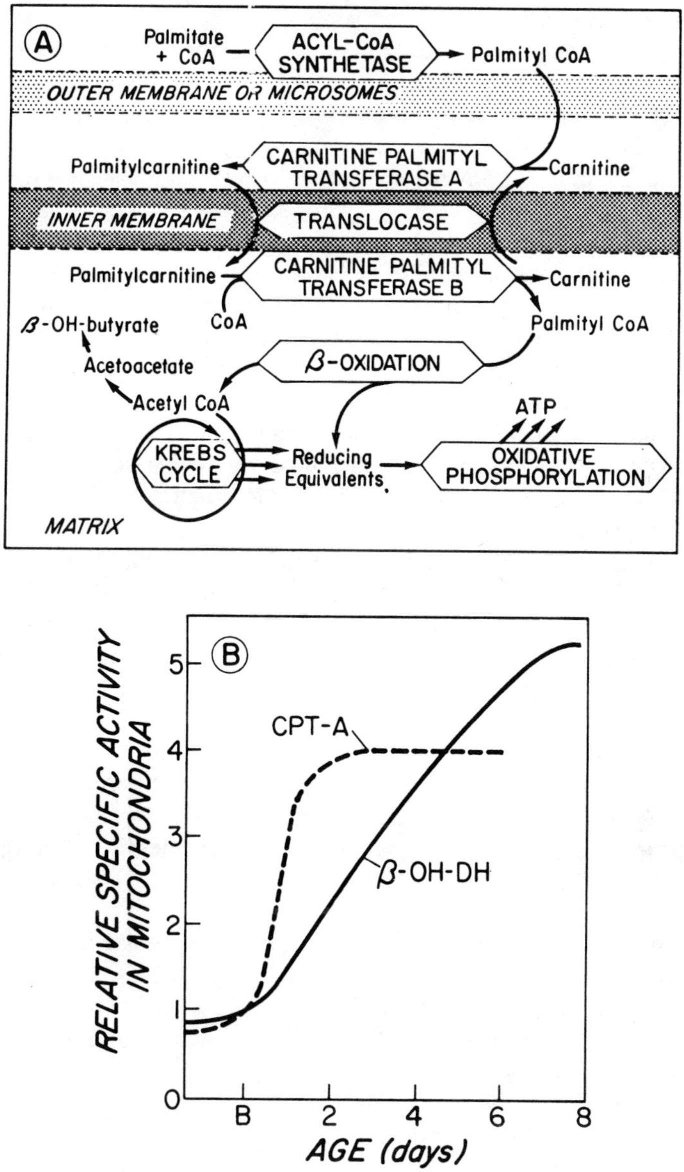

Figure 3-5. Subcellular compartmentation of the long chain fatty acid oxidation pathway (panel *A*), and development of carnitine palmitoyl transferase A (CPT) and β-hydroxylbutyrate dehydrogenase (β-OH-DH) (panel *B*).

fatty acid oxidation has other roles such as promoting gluconeogenesis by maintaining a high redox state (Böhme, Sparmann, and Hofmann, 1983; Girard et al., 1981; El Manoubi, et al., 1983) and producing ketone bodies, an important fuel source for neonatal brain (Hawkins, Williamson, and Krebs, 1971).

With respect to the importance of ketone body production, it is significant that the activity β-OH butyrate dehydrogenase (a mitochondrial enzyme) develops rapidly in the postnatal period (Knowles and Ballard, 1974; Levy and Toury, 1970; Hommes, Luit-de Hann, and Richters, 1971; Hallman, 1971*a*) on a schedule following the development of carnitine acyltransferase activity (Fig. 3-5*B*). The postnatal emergence of these two enzymes is a clear example of mitochondrial differentiation, in this case a specialization for processing the ketogenic diet of the newborn. In the whole tissue the metabolic consequences of the enzyme increases in mitochondria are magnified by mitochondrial proliferation during early postnatal life.

Panel *B* in Figure 3-5 summarizes the developmental profile for carnitine palmityltransferase A (— —) (Bailey, 1981; Chalk et al., 1983; Rulfs, 1981) and β-OH-butyrate dehydrogenase (Levy and Toury, 1970; Hommes, Luit-de Hann, and Richters, 1971; Hallman, 1971*a*) in isolated rat liver mitochondria. The developmental patterns are composites derived from activities reported in the references just cited. Specific activities for isolated mitochondria are plotted relative to their values at birth, set arbitrarily at 1.0.

Several intramitochondrial enzymes serve important functions in ureogenesis. Two of these enzymes, carbamoyl phosphate synthetase and ornithine transcarbamylase, are part of the five-enzyme sequence comprising the urea cycle. Their combined activities in the mitochondria synthesize citrulline from ammonia, ornithine and HCO_3^- in an ATP-dependent reaction sequence (Fig. 3-6*A*). Glutamate dehydrogenase (already discussed) is also important because it has a central position in the catabolism of amino acids and supplies ammonia for the carbamoyl phosphate synthetase reaction (Krebs et al., 1978). N-acetylglutamate synthetase synthesizes a required activator of carbamoyl phosphate synthetase from acetyl CoA and glutamate. The matrix n-acetylglutamate synthetase develops slowly in the postnatal period (Rabier et al., 1983) and will not be discussed further here.

The five enzyme activities of the urea cycle begin to develop together during the last three days of gestation except that argininosuccinate synthetase lags a bit and seems to be rate-limiting for urea synthesis (Fig. 3-6*B* inset) (Räihä and Suihkonen, 1968; Miller and Chu, 1970). Postnatally, mitochondria differentiate rapidly with respect to carbamoyl phosphate synthetase but the specific activity of ornithine transcarbamylase in isolated mitochondria does not increase nearly as much (Fig. 3-6*B*) (Rabier et al., 1983; Murakami, Kitagawa, and Sugimoto, 1983). The postnatal increase in ornithine

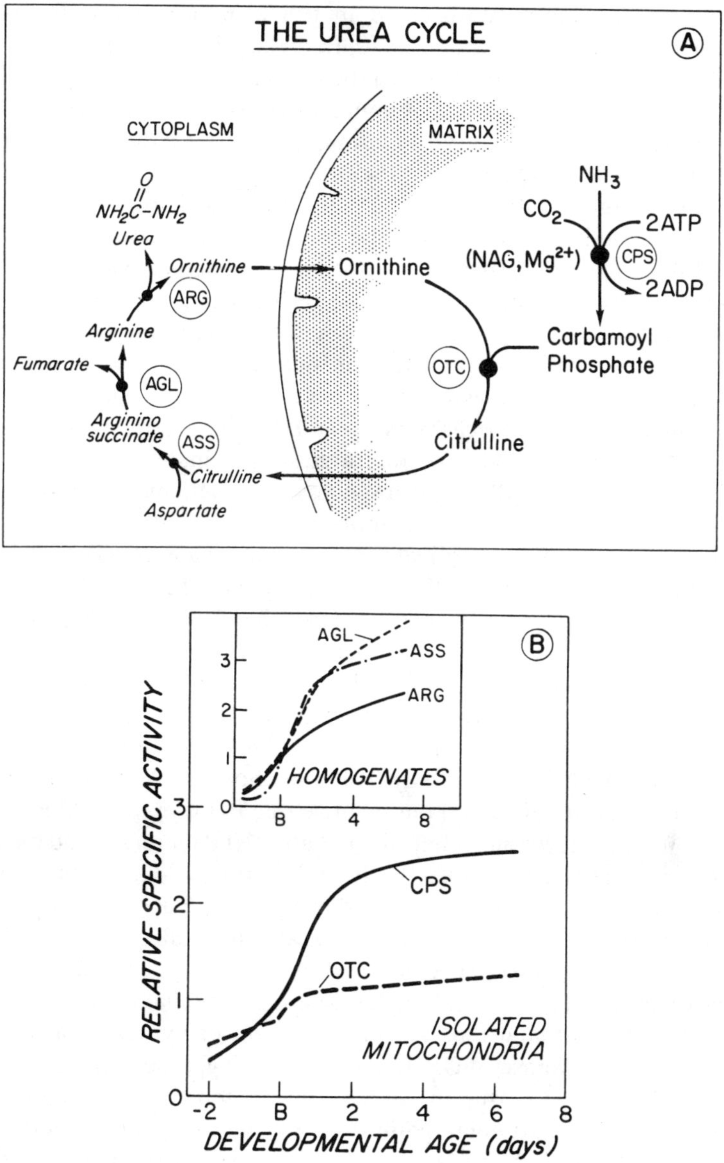

Figure 3-6. Subcellular compartmentation and development of urea cycle enzymes.

transcarbamylase activity observed in tissue homogenates (Räihä and Suihkonen, 1968; Miller and Chu, 1970; Belbekouche, Gautier, and Vaillant, 1985; Gautier and Vaillant, 1979) can be attributed almost entirely to mitochondrial proliferation. A pool of immunoreactive but enzymatically inactive ornithine transcarbamylase accumulates during late gestation and is then assembled and/or activated after birth in mitochondria during the most rapid phase of proliferation (Fig. 3-6*B*) (Belbekouche, Gautier, and Vaillant, 1985). Mitochondrial proliferation also has the effect of amplifying the differentiative increase in carbamoyl phosphate synthetase activity. It is interesting that the development of the matrix enzymes occurs in parallel with the development of the cytoplasmic enzymes of the urea cycle (Fig. 3-6*B*).

In Figure 3-6 panel *A* is a diagram of the urea cycle showing the matrix reactions of citrulline synthesis catalyzed by carbamoyl phosphate synthetase (CPS) and ornithine transcarbamylase (OTC). The remaining three enzymes of the cycle are cytoplasmic: argininosuccinate synthetase (ASS), argininosuccinate lyase (AGL, or argininosuccinase), and arginase (ARG). The CPS reaction is diagrammed in more detail to show its ATP dependence and the requirement for the activator n-acetylglutamate (NAG), both important factors in postnatal regulation of citrulline synthesis. Panel *B* shows the development of CPS and OTC measured in isolated mitochondria (Rabier et al., 1983); the inset shows ASS, AGL, and ARG measured in whole homogenates (Räihä and Suihkonen, 1968; Miller and Chu, 1970). Specific activities are plotted relative to an arbitrary value of 1.0 at birth.

Carbamoyl phosphate synthetase is probably rate-limiting for citrulline synthesis in intact isolated newborn liver mitochondria. At birth this enzyme is present (Rabier et al., 1983) but inactive due to low matrix ATP concentrations. Within a few hours after birth, matrix ATP increases and citrulline synthesis increases several fold (Kelly and Aprille, 1984). Then for about four days it is the concentration of the activator n-acetylglutamate that appears to limit full expression of enzyme activity (Rabier et al., 1983).

In summary, several key special function enzymes emerge in liver during the active phases of mitochondrial development that characterize late gestation and early postnatal life. The development of pyruvate carboxylase, carnitine acyltransferase, β-OH butyrate dehydrogenase, carbamoyl phosphate synthetase, and ornithine transcarbamylase is coordinated with the emergence of the particular pathways of which they are a part. Their insertion into the mitochondria is clearly an example of biochemical differentiation of the organelle, superimposed over the structural differentiation that occurs prenatally and the proliferation that occurs postnatally. It may be coincidence that these pathways of intermediary metabolism develop at the time of most active mitochondrial biogenesis in the liver. On the other hand there is the intriguing possibility that mitochondrial events control or are controlled by some aspect of nuclear gene expression necessary for the coordinated synthesis of pathways that have some enzymes destined for the mitochondria. At the very least, the fact that so many major pathways have at least one enzyme localized in the mitochondrion points to the importance of this organelle as the main intersection for chemical traffic in the cell. This localization provides great potential for integrated regulation of metabolic fluxes, especially since the interactive enzymes are contained in a relatively small subcellular compartment that is selectively permeable to specific substrates and products.

INTEGRATED FUNCTIONS OF INTACT LIVER MITOCHONDRIA

Respiratory Control in Fetal Liver Mitochondria

It is important to dispel the notion that fetal liver mitochondria (18 days gestation to birth) are "leaky" to protons or "loosely coupled" with respect to oxidative phosphorylation. This conclusion was reached in a number of early studies because criteria for assessing coupling efficiency that had been established in adult mitochondria were applied to the fetal liver and misinterpreted. Normal adult liver mitochondria respire slowly in the presence of an oxidizable substrate, but oxygen consumption is greatly stimulated upon addition of ADP. This stimulation is called *acceptor control* because the phosphorylation of ADP is the acceptor for P_i and for the energy of the proton gradient established by substrate oxidation. The ratio of the ADP-stimulated respiratory rate (state 3) to the slower rate prior to ADP addition is called the acceptor control ratio (ACR). Once all the added ADP has been phosphorylated, a slower rate of respiration ensues (state 4). The ratio of state 3 to state 4 is called the respiratory control index or the respiratory control ratio (RCR).

In early studies of fetal mitochondria it was noticed that both the ACR and RCR were low relative to values obtained for well-coupled adult mitochondria (Pollak and Duck-Chong, 1973; Mintz et al., 1967; Hallman, 1971*a;* Nakazawa et al., 1973; Pollak, 1977). These findings were often interpreted (Pollak and Duck-Chong, 1973) to mean that fetal mitochondria were loosely coupled, implying that they oxidized substrates at a fast rate with inefficient use of the available potential energy for ATP synthesis. In adult mitochondria low control ratios typically result when the denominators are large, that is, when respiration in the absence of ADP is unusually fast. Re-examination of this problem in late fetal mitochondria showed that the rate of substrate oxidation in the absence of ADP was no greater than in adult mitochondria (Aprille and Asimakis, 1980). The low control ratios are instead due to an unusually low state 3 rate (the numerator of the control ratios) (Fig. 3-7). The evidence pointed not to leaky or uncoupled mitochondria, but to well-coupled oxidative phosphorylation that seemed to be supressed by a novel control phenomenon. This conclusion was suggested in 1963 by Van Rossum from studies of oxygen consumption in liver slices but was not pursued. A key observation in both Van Rossum's report and recent work (Nakazawa et al., 1973; Aprille and Asimakis, 1980) was that respiratory rates were greatly stimulated by uncouplers (Fig. 3-7). This respiratory stimulation would not be possible if the mitochondria were already inherently uncoupled.

In summary, oxidation and phosphorylation in fetal liver mitochondria are probably as well coupled as in mitochondria from postnatal livers. Fetal mitochondria do have lower state 3 rates, and though they must synthesize ATP at a slower rate they do not operate with energetic inefficiency. There is

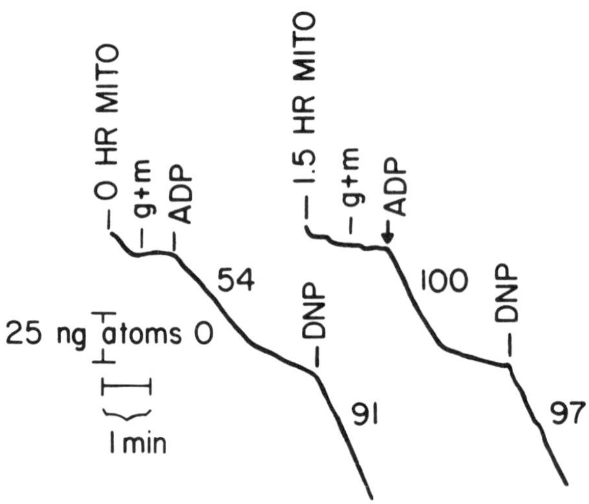

Figure 3-7. Polarographic measurement of respiratory rates in liver mitochondria from newborns (left) and from animals 1.5 hours old (right). Additions were 5 mM each glutamate + malate (g + m), 150 nmoles ADP to elicit state 3, and 40 μM 2,4-dinitrophenol (DNP) to elicit uncoupled rates. Note that state 3 is much lower at birth than at 1.5 hours (54 vs 100 ngatoms O/min/mg protein respectively) whereas state 4 and uncoupled rates are similar (from Aprille and Asimakis, 1980).

no evidence for impaired energy transduction or for wasted chemical energy due to loose coupling.

State 3 Respiration in Relation to the Matrix Adenine Nucleotide Pool Size

The theories originally put forward to explain the perinatal development of the RCR were based on the erroneous assumption that the development of the RCR was due to a decrease in the membrane permeability to protons, thought to be linked to perinatal change in the morphological and physical characteristics of mitochondrial membranes (Pollak and Duck-Chong, 1973; Pollak and Munn, 1970; Pollak, 1975, 1977; Sandor and Pollak, 1976; Sutton and Pollak, 1978, 1980). Adenine nucleotides were suspected to play a role, possibly by initiating conformational changes in the mitochondria (Pollak, 1977; Sandor and Pollak, 1976; Sutton and Pollak, 1978, 1980). This impression was enhanced by two findings: the mitochondria have a low adenine nucleotide content that increases to adult levels shortly after birth (Nakazawa et al., 1973; Aprille and Asimakis, 1980; Sutton and Pollak, 1978) and energy-linked functions were improved by a pre-incubation with ATP and Mg^{2+} (Mintz et al., 1967; Hallman, 1971*a;* Nakazawa et al., 1973; Sandor and Pollak, 1976; Bygrave and Ash, 1977). Once it became clear that low control ratios were due to slow state 3 respiration rather than fast state 4 respiration (Nakazawa et al., 1973; Aprille and Asimakis, 1980), state 3 was discovered to be directly limited by the very low matrix adenine nucleotide pool size (Aprille and Asimakis, 1980; Asimakis and Aprille, 1980*a*).

Several viable hypotheses have been suggested to explain the regulatory effect of matrix adenine nucleotide content on state 3 respiration. The rate of ADP-ATP exchange over the adenine nucleotide translocase is a linear function of the ATP + ADP pool size probably because substrate concentration on the matrix side is limiting for carrier activity (Nakazawa et al., 1973; Aprille and Asimakis, 1980; Asimakis and Aprille 1980*a;* Hale and Williamson, 1984; Rulfs and Aprille, 1982). Given the exceptionally low matrix pool size found in fetal rat liver mitochondria, translocase activity may limit the availability of ADP for phosphorylation, at least under conditions that pertain in vitro (Aprille and Asimakis, 1980; see also Asimakis and Aprille, 1980*a;* Hale and Williamson, 1984; Rulfs and Aprille, 1982; Baggeto et al., 1984). With respiratory control intact, the low translocase activity limits the rate of state 3 respiration. However, control strength studies show the ATPase-ATP synthetase complex, not the adenine nucleotide translocase, to be an important rate-controlling point in the newborn (Baggeto, Gautheron, and Godinot (1984). Possibly the low absolute matrix concentration of ADP, a substrate for the ATP synthetase, may be submaximal relative to the in situ K_m for the phosphorylating reaction (Nakazawa et al., 1973; see also discussion in Aprille and Asimakis, 1980). Other suggested alternatives are that the matrix adenine nucleotide concentration affects the conformation of the inner membrane (Sutton and Pollak, 1980) or activates some specific protein required for maximum rates of oxidative phosphorylation (Baggeto et al., 1983; Baggeto, Gautheron, and Godinot, 1984).

Within minutes after birth, the mitochondrial adenine nucleotide pool size begins to increase (Fig. 3-8) (Nakazawa et al., 1973; Aprille and Asimakis, 1980; Sutton and Pollak, 1978). This increase must occur by accumulation from the cytoplasm since the total adenine nucleotide content of the cell does not change (Sutton and Pollak, 1978; Van Lelyveld and Hommes, 1978; Cuezva and Medina, 1982; Ballard, 1970; Cuezva, Chitra, and Patel, 1982). By one hour, the matrix pool size has increased to about 5 to 6 nmol/mg protein, which is sufficient to support maximal rates of state 3 respiration (Fig. 3-8) (Aprille and Asimakis, 1980; Asimakis and Aprille, 1980*a*). Translocase activity develops as a linear function of the matrix ATP + ADP pool size. The RCR also increases along with state 3, of course, but then remains relatively stable throughout postnatal development (Nakazawa et al., 1973; Aprille and Asimakis, 1980). ADP/O ratios are hard to measure polarographically in fetal mitochondria because of the indefinite transition from state 3 to state 4 (Fig. 3-7), but most studies report ADP/O ratios (or P/O ratios) near the theoretical values (Mintz et al., 1967; Hallman, 1971*a;* Dawkins, 1959; Miyahara et al., 1984; Aprille and Asimakis, 1980; Wilson, 1972; Holtzman and Moore, 1973).

A cause-effect relationship between the increased matrix pool size and development of state 3 rates is indicated by the following experimental evidence. Isolated term liver mitochondria will take up adenine nucleotides in vitro, and the translocase rate, state 3, and RCR increase as they do during in vivo development (Nakazawa et al., 1973; Pollak, 1977; Aprille and Asimakis, 1980; Sandor and Pollak, 1976; Aprille, 1981). Conversely, if adult liver mitochondria are depleted of adenine nucleotides in vitro, state 3 respiration and the RCR are predictably decreased when the pool size falls below 5 to 6 nmol/mg protein, and translocase decreases as a linear function of matrix adenine nucleotide content (Asimakis and Aprille, 1980*a*).

In 1963 Van Rossum observed that oxygen consumption in fetal rat slices declines during late gestation, reaches a nadir at term, then rapidly increases again within one day of birth. In the presence of an uncoupler, oxygen consumption remained constant as a function of age, prompting Van Rossum to conclude that there must be some unusual control of coupled respiration in the perinatal period. The findings described here show that his interpretation was correct.

Postnatal Adenine Nucleotide Accumulation in Mitochondria

The ATP + ADP + AMP content of mitochondria continues to increase for about four hours after birth, well beyond what is necessary for full development of the state 3 respiratory rate. This postnatal phenomenon has been observed in the rabbit (Rulfs and Aprille, 1982) and guinea pig (Hale and Williamson, 1984) as well as the rat. In general, the rapid and extensive postnatal accumulation of adenine nucleotides by liver mitochondria raises three very interesting questions: (1) What is the transport mechanism that allows this to happen, since the only previously known nucleotide carrier operates as a one-for-one exchange? (2) What stimulus or change in cellular status at the time of birth signals the transport process to begin? (3) Besides promoting the rapid maturation of state 3 respiration, what other physiological significance might be attached to the rapid increase in the matrix ATP, ADP, and AMP concentrations? Progress made toward answering these questions will be reviewed briefly.

Net Transport of Adenine Nucleotides. The transport mechanism for the net accumulation of adenine nucleotides was initially thought to involve the ADP-ATP exchange translocase because this carrier activity increases postnatally (Nakazawa et al., 1973). The development of translocase activity was subsequently shown to be a consequence, not a cause, of the increased adenine nucleotide pool size (Aprille and Asimakis, 1980; Asimakis and Aprille, 1980*a;* Rulfs and Aprille, 1982; Aprille, 1981). Kinetic properties of the translocase remain constant during development (Pollak and Sutton, 1978), and because this carrier operates as a one-for-one exchange (Klingenberg, 1980), it cannot cause net accumulation of adenine nucleotides. Further study revealed another transport process for adenine nucleotides that was much slower, Mg^{2+}-dependent, ATP-specific, and carboxyatractyloside-insensitive, and therefore distinct from the exchange translocase (Aprille and Asimakis, 1980; Aprille, 1981; Pollak and Sutton, 1980*b,* 1980*c*).

The new transport activity seems to be a general feature of adult as well as newborn liver mitochondria (Austin and Aprille, 1983, 1984; Aprille and Austin, 1981). The carrier appears to catalyze the coupled exchange of external ATP · Mg with matrix P_i to bring about a net increase in the matrix adenine nucleotide pool size (Austin and Aprille, 1984). There is normally a P_i concentration gradient across the inner membrane (high inside) in proportion to the ΔpH. ATP · Mg probably exchanges for matrix P_i until the adenine nucleotide concentration gradient approximates that of the P_i gradient, which remains constant via the much faster P_i^-/OH^- carrier. At this point a steady state is reached in which exogenous ATP · Mg is as likely to exchange for matrix adenine nucleotides as for matrix P_i, so the adenine nucleotide pool size also remains constant. The relationship with the P_i gradient makes net accumulation indirectly dependent on ΔpH (Aprille and Austin, 1981). The exact stoichiometry, total energy requirements, and charge balance for this transporter remain to be worked out, but several other kinetic properties have been described (Aprille, 1981; Pollak and Sutton, 1980*b;* Aprille and Austin, 1981; Asimakis and Aprille, 1980*b*). For instance, ATP · Mg is preferred over ADP · Mg as a substrate. Therefore, the rate of transport and the direction of net flux of adenine nucleotides depends on the ATP/ADP ratios on both sides of the membrane as well as on the P_i concentration gradient (Austin and Aprille, 1983, 1984).

Postnatal Regulation of Transport. Postnatal adenine nucleotide accumulation in vivo must be a consequence of change in the cellular milieu at parturition, because isolated fetal or newborn liver mitochondria maintain a normal P_i pool size (Nakazawa et al., 1973; Baggeto et al., 1983) and a protonmotive force (Prpic and Bygrave, 1981) equivalent to adult mitochondria and will accumulate adenine nucleotides in vitro (Nakazawa et al., 1973; Aprille and Asimakis, 1980; Sutton and Pollak, 1980; Aprille, 1981; Pollak and Sutton, 1980*b*). The requirements for transport suggest circumstances that might account for adenine nucleotide accumulation immediately after birth. The ATP/ADP ratio of the liver cytoplasm increases several fold within minutes after parturition (Knowles and Ballard, 1974; Ballard, 1971; Sutton and Pollak, 1978; Van Lelyveld and Hommes, 1978; Cuezva and Medina, 1982; Ballard, 1970) producing a large concentration gradient across the inner mitochondrial membrane (high outside) for ATP · Mg, the preferred transport substrate. Also, the liver is relatively hypoxic during late gestation, but becomes normoxic a short time after delivery with the onset of independent breathing and the perinatal change in systemic circulation. This change in oxygenation could promote a higher energy state in the mitochondrion, a prerequisite for the P_i gradient that in turn is necessary for adenine nucleotide accumulation against a concentration gradient. If rats or rabbits are kept under hypoxic conditions after birth, the uptake of adenine nucleotides by liver mitochondria is delayed (Sutton and Pollak, 1980; Aprille, Nosek, and Brown, 1983; Brennan and Aprille, 1983; Brennan and Aprille, 1985), perhaps because under these circumstances oxygenation is inadequate for full development of the protonmotive force. In adult livers, the matrix adenine nucleotide pool becomes depleted if livers are subjected to hypoxic perfusion (Watanabe et al., 1983). Oxygenation alone is not a sufficient condition for adenine nucleotide uptake, however, because uptake into mitochondria does not occur in isolated fetal liver cells incubated in vitro under oxygen (Van Lelyveld and Hommes, 1978).

The hormone status of the newborn probably also modulates the rate of adenine nucleotide accumulation either directly, or indirectly through changes in the cytoplasmic ATP/ADP. Catecholamines, possibly through β-adrenergic action, may be important (Sutton and Pollak, 1980). Glucagon or the insulin/glucagon ratio is probably also regulatory (Meister et al., 1983; Sutton and Pollak, 1978, 1980; Aprille, Nosek, and Brown, 1983; Brennan and Aprille, 1985). The normal postnatal decline in insulin/glucagon is delayed in rat pups born of diabetic mothers (Cuezva et al., 1982) as is the normal increase in liver mitochondrial adenine nucleotides (Aprille, Nosek, and Brown, 1983). Glucose injection delays the postnatal maturation of state 3 respiration (Meister et al., 1983), probably by increasing the insulin/glucagon ratio. The exact mechanism by which altered ATP/ADP ratios, tissue oxygenation, and hormones combine to promote the net uptake of adenine nucleotides into the mitochondria remains to be determined, but all of these factors appear to be important.

Physiological Significance. The magnitude of the postnatal increase in matrix ATP, ADP, and AMP is much greater than what is necessary for the development of state 3 respiration (Fig. 3-8). The physiological significance of this phenomenon in relation to other mitochondrial functions in the liver has just begun to be explored. The approach has been to consider which mitochondrial reactions could be affected by an increase in the matrix adenine nucleotide concentration. Pyruvate carboxylase is an obvious candidate. It requires ATP for the conversion of pyruvate to oxaloacetic acid, the first step in gluconeogenesis (Fig. 3-4*A*). This reaction is stimulated by the postnatal increase in the matrix ATP concentration in both the rat (Aprille, Yaswen, and Rulfs, 1981) and the rabbit (Brennan and Aprille, 1984*a*, 1984*b*). The sudden activation of pyruvate carboxylase is responsible for the initial onset of gluconeogenesis in the newborn, contributing to rapid recovery from transient postnatal hypoglycemia (Brennan and Aprille, 1983, 1984*a*, 1984*b*, 1985). Any cause of delay in the postnatal uptake of adenine nucleotides may also delay the activation of gluconeogenesis, perhaps accounting for the prolonged hypoglycemia characteristic of infants with respiratory distress (hypoxia) or from diabetic mothers (Aprille, Nosek, and Brown, 1983; Brennan and Aprille, 1983, 1984*b*, 1985).

There are a number of reported observations that may be explained by the postnatal increase in matrix adenine nucleotides. For example, uncoupler-stimulated F_0F_1ATPase activity in intact mitochondria is low at birth and develops postnatally, but if activity is measured in disrupted mitochondria, there is no change with developmental age (Hallman, 1971*a*). This difference is probably due to the rate-limiting concentration of matrix ATP, which is submaximal for either F_0F_1ATPase or translocase activity, a necessary intermediate step in the reaction. Another example is the somewhat anomalous regulation of pyruvate dehydrogenase activity with respect to ATP/ADP ratios in fetal liver mitochondria (Knowles and Ballard, 1974). When the matrix ATP/ADP is high, pyruvate dehydrogenase is unexpectedly active (Berger and Hommes, 1975), perhaps explainable if the absolute concentrations of ATP and ADP are considered along with the ATP/ADP ratio. Maybe the concentrations of ATP and ADP in the matrix are too low either to exert a regulatory effect on the pyruvate dehydrogenase kinase or to provide sufficient ATP for the kinase reaction. After birth, normal regulation comes into play as adenine nucleotides are accumulated and the matrix concentration of ATP and ADP increase. This scenario may be important for a switch in pyruvate flux from oxidation to the pyruvate carboxylase reaction and gluconeogenesis, but it remains to be proven.

Many other matrix reactions may be regulated by the rapid postnatal increase in the adenine pool size. Preliminary data suggest that carbamoyl phosphate synthetase activity is stimulated by the accumulation of ATP,

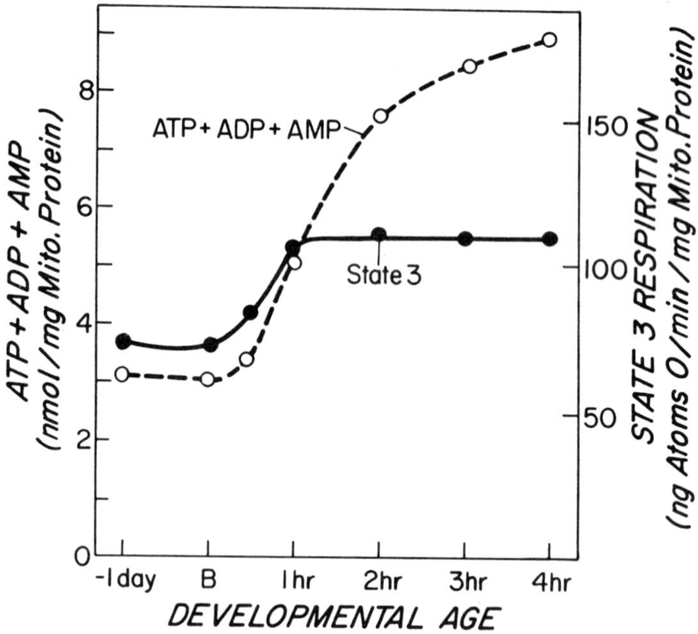

Figure 3-8. Adenine nucleotide content and state 3 respiration in rat liver mitochondria during development. State 3 respiration was determined as described in Figure 3-7. The ATP + ADP + AMP content was determined enzymatically in perchloric acid extracts of isolated mitochondria as described by Aprille and Asimakis (1980).

resulting in increased rates of citrulline synthesis soon after birth (Kelley and Aprille, 1984). Ca^{2+} transport normally develops within two hours after birth; it is induced prematurely by glucagon and is promoted in vitro by incubation of fetal liver mitochondria with ATP · Mg (Bygrave and Ash, 1977; Prpic and Bygrave, 1981). The ATP-dependent formation of acylCoA esters of many organic acids is another likely candidate for this type of regulation. The kinase-phosphatase cycle of branched chain α-keto-acid dehydrogenase regulation (Paul and Adibi, 1982) may be affected by ATP and ADP concentrations in the matrix in a manner analogous to that proposed above for pyruvate dehydrogenase. Matrix enzymes that are allosterically regulated by ATP, ADP, or AMP might be susceptible to rapid changes in the adenine nucleotide pool size as well as to the ATP/ADP ratio.

Perhaps most interesting is the prospect that the intramitochondrial reactions of organelle biogenesis might be regulated by adenine nucleotides. Certainly the intramitochondrial reactions of DNA replication, transcription, and translation, and the energy-dependent uptake of mitochondrial proteins synthesized in the cytoplasm might be affected by the concentration of ATP in the matrix compartment. The postnatal increase in carbamoylphosphate synthetase and ornithine transcarbamoylase activities is clearly triggered by some factor associated with birth, and is repressed by glucose injection but not by actinomycin D (Murakami, Kitagawa, and Sugimoto, 1983; Gautier and Vaillant, 1979; Belbekouche, Gautier, and Vaillant, 1985). The specific activity of in vivo ^{3}H-leucine incorporation into mitochondrial subpopulations differs in immature and mature liver (Pollak, 1976). Mitochondrial protein synthesis measured in vitro declines during late gestation, then develops rapidly after birth to a peak within 12 hours (Hallman, 1971*a*). Furthermore, the postnatal development of mitochondrial protein synthesis is delayed if the animals are kept in hypoxic conditions for 12 hours from birth but not if they are allowed 1 hour of normoxia first (Hallman, 1971*a*). This difference seems likely to be related to the hypoxia-delayed increase in matrix adenine nucleotide concentration previously noted. The normal proliferation of mitochondria measured as mg mitochondrial protein yield/g tissue is also suppressed by hypoxia (Table 1 in Kinnula and Hassinen, 1977) as is the postnatal increase in cytochrome aa_3/mg mitochondrial protein (Kinnula and Hassinen, 1977). A comparison of Figures 3-1 and 3-8 shows that the enlargement of mitochondria during late gestation is correlated with a very low matrix adenine nucleotide pool size. Within a very short time after birth, mitochondria rapidly proliferate and differentiate on a schedule that suggests the accumulation of adenine nucleotides might be permissive or even be the triggering stimulus. This situation promises to be a fruitful area of further study.

CONCLUSIONS

The complex series of developmental changes that have been reviewed here seem to resolve into three distinct phases of mitochondrial development. During the first phase, which covers the final three to four days of gestation, mitochondria differentiate with respect to enzyme composition and morphology, but functionally they seem to be relatively dormant. Parenchymal cell mitochondria are larger in size and metabolically less active just before birth than at any other time in their natural history. The second phase of mitochondrial development occurs in the first few hours after birth. The rapid accumulation of adenine nucleotides into existing mitochondria results in the full development of bioenergetic function and activates the metabolic activities that are minimally necessary to ensure survival during the transition to extrauterine life. This phenomenon also seems to set the stage for the final phase of postnatal metabolic adaptation that occurs during the next several days. The activated mitochondria begin to proliferate and continue to differentiate,with the acquisition of newly synthesized enzymes as the predominant feature of the further development of metabolic pathways such as gluconeogenesis, fatty acid oxidation, and urea synthesis.

In many respects the second phase of development is the most interesting because very dramatic changes occur rapidly with profound consequences for successive developmental events. The elucidation of the initiating stimuli and the biochemical mechanisms responsible for these rapid metabolic adjustments that occur in liver mitochondria at birth is an intriguing challenge for future investigation.

ACKNOWLEDGMENTS

Support for this project was provided by NIH grants HD16936 and NS14936, the Marion and Jasper Whiting Foundation, and Tufts University. The author is indebted to Dr. William A. Brennan and Dr. Gary Fiskum for critical editing, and to Ms. Suzanne Dobson for excellent secretarial assistance.

REFERENCES

Aprille, J. R., 1981, Net uptake of adenine nucleotides by newborn rat liver mitochondria, *Archiv. Biochem. Biophys.* **207**:157-164.

Aprille, J. R., and G. K. Asimakis, 1980, Postnatal development of rat liver mitochondria: State 3 respiration, adenine nucleotide translocase activity, and the net accumulation of adenine nucleotides, *Archiv. Biochem. Biophys.* **201**:564-75.

Aprille, J. R., and J. Austin, 1981, Regulation of the mitochondrial adenine nucleotide pool size, *Archiv. Biochem. Biophys.* **212**:689-699.

Aprille, J. R., M. T. Nosek, and C. A. Brown, 1983, Delayed postnatal development of rat liver mitochondria in hypoxia and in newborns from diabetic mothers, *Pediatr. Res.* **17**:126A.

Aprille, J. R., P. Yaswen, and J. Rulfs, 1981, Acute postnatal regulation of pyruvate carboxylase activity by compartmentation of mitochondrial adenine nucleotides, *Biochim. Biophys. Acta* **675:**143-147.

Asimakis, G. K., and J. R. Aprille, 1980*a*, *In vitro* alteration of the size of the liver mitochondrial adenine nucleotide pool: Correlation with respiratory functions, *Archiv. Biochem. Biophys.* **203:**307-316.

Asimakis, G. K., and J. R. Aprille, 1980*b*, Net uptake of adenine nucleotides in isolated rat liver mitochondria, *FEBS Lett.* **117:**157-160.

Austin, J., and J. R. Aprille, 1983, Net adenine nucleotide transport in rat liver mitochondria is affected by both the matrix and the external ATP/ADP ratios, *Archiv. Biochem. Biophys.* **222:**321-325.

Austin, J., and J. R. Aprille, 1984, Carboxyatractyloside-insensitive influx and efflux of adenine nucleotides in rat liver mitochondria, *J. Biol. Chem.* **259:**154-160.

Baggeto, L., D. C. Gautheron, and C. Godinot, 1984, Effects of ATP on various steps controlling the rate of oxidative phosphorylation in newborn rat liver mitochondria, *Archiv. Biochem. Biophys.* **232:**670-678.

Baggeto, L., J. Comte, R. Meister, and C. Godinot, 1983, High phosphate requirement for oxidative phosphorylation and low affinity for phosphate transport in newborn rat liver mitochondria, *Biochimie* **65:**685-690.

Bailey, E., 1981, Regulation of hepatic lipid metabolism during development of the rat, *Biochem. Soc. Trans.* **9:**371-372.

Ballard, F. J., 1970, Adenine nucleotides and the adenylate kinase equilibrium in livers of foetal and newborn rats, *Biochem. J.* **117:**231-235.

Ballard, F. J., 1971, The development of gluconeogenesis in rat liver. Controlling factors in the newborn, *Biochem. J.* **124:**265-274.

Ballard, F. J., and R. W. Hanson, 1967, Phosphoenolpyruvate carboxykinase and pyruvate carboxylase in developing rat liver, *Biochem. J.* **104:**866-871.

Belbekouche, M., C. Gautier, and R. Vaillant, 1985, Changes in ornithine transcarbamylase activity and protein level during perinatal period in rat liver. Effects of actinomycin D, *Biochem. Biophys. Res. Commun.* **129:**780-788.

Berger, R., and F. A. Hommes, 1975, Regulation of pyruvate metabolism in fetal rat liver, in *Normal and Pathological Development of Energy Metabolism*, F. A. Hommes and C. J. Van den Berg, eds., Academic Press, London, pp. 97-107.

Bieber, L. L., T. Helmrath, E. A. Dolanski, L. L. Belanger, and Y. R. Choi, 1980, Studies on the formation of phosphoenolpyruvate and citric acid cycle intermediates by neonatal piglet liver mitochondria, *Biol. Neonate* **37:**197-203.

Bittner, R., H. J. Böhme, L. Didt, W. Goitzsch, E. Hofmann, M. J. Levin, and G. Sparmann, 1979, Developmental changes in the levels of hepatic enzymes and their relation to metabolic functions, *Adv. Enz. Reg.* **17:**37-57.

Böhme, H. J., G.. Sparmann, and E. Hofmann, 1983, Biochemistry of liver development in the perinatal period, *Experentia* **39:**473-483.

Brennan, W. A., Jr., and J. R. Aprille, 1983, Hypoxia retards the development of mitochondrial pyruvate carboxylation in newborn rabbit liver, *Pediatr. Res.* **17:**126.

Brennan, W. A., Jr., and J. R. Aprille, 1984*a*, Regulation of hepatic gluconeogenesis by rapid compartmentation of mitochondrial adenine nucleotides in the newborn rabbit, *Comp. Biochem. Physiol.* **77B:**35-39.

Brennan, W. A., Jr., and J. R. Aprille, 1984*b*, Pyruvate carboxylase and the acute regulation of hepatic gluconeogenesis in the newborn rabbit, *Fed. Proc.* **43:**2003.

Brennan, W. A., Jr., and J. R. Aprille, 1985, Regulation of hepatic gluconeogenesis in newborn rabbit: Controlling factors in presuckling period, *Am. J. Physiol.* **249:**E498-E505.

Bygrave, F. L., and G. R. Ash, 1977, Development of mitochondrial calcium transport activity in rat liver, *FEBS Lett.* **80:**271-274.

Chalk, P. A., F. C. Higham, A. M. Caswell, E. Bailey, 1983, Hepatic mitochondrial fatty acid oxidation during the perinatal period in the rat, *Internat. J. Biochem.* **15:**531-538.

Chang, L. O., 1977, The development of pyruvate carboxylase in rat liver mitochondria, *Pediatr. Res.* **11:**6-8.

Chen, L. B., I. C. Summerhayes, L. V. Johnson, M. L. Walsh, S. D. Bernal, and T. J. Lampidis, 1982, Probing mitochondria in living cells with rhodamine 123, *Cold Spring Harbor Symp. Quant. Biol.* **46:**141-155.

Cuezva, J. M., C. I. Chitra, and M. S. Patel, 1982, The newborn of diabetic rat. II. Impaired gluconeogenesis in the postnatal period, *Pediatr. Res.* **16:**638-643.

Cuezva, J. M., and J. M. Medina, 1982, Adenine nucleotide concentrations in liver of fetal rats. Neonatal changes in the premature newborn, *Rev. Esp. Fisiol.* **38:**161-166.

Cuezva, J. M., E. S. Burkett, D. S. Kerr, H. M. Rodman, and M. S. Patel, 1982, The newborn of diabetic rat. I. Hormonal and metabolic changes in the postnatal period, *Pediatr. Res.* **16:**632-637.

Dallman, P. R., and H. C. Schwartz, 1964, Cytochrome c concentrations during rat and guinea pig development, *Pediatrics* **33:**106-110.

David, H., 1979, Quantitative and qualitative changes in the mitochondria in hepatocytes during postnatal development of male rats, *Exp. Pathol.* **17:**359-373.

Dawkins, M. J. R., 1959, Respiratory enzymes in the liver of the newborn rat, *R. Soc. London Proc.*, Ser. B, **150:**284-298.

Dawkins, M. J. R., 1966, Biochemical aspects of developing function in newborn mammalian liver, *Brit. Biol. Bull.* **22:**27-33.

DeVos, M. A., C. W. Wilmink, and F. A. Hommes, 1968, Development of some mitochondrial oxidase systems of rat liver, *Biol. Neonate* **13:**83-89.

El Manoubi, L., S. Callikan, P. H. Duee, P. Ferre, and J. Girard, 1983, Development of gluconeogenesis in isolated hepatocytes from the rabbit, *Am. J. Physiol.* **244:**E24-E30.

Escriva, F., J. F. Decaux, P. Ferre, and J. R. Girard, 1984, Evidence that hepatic mitochondrial mass decreases during the first sixteen hours following birth in starved newborn rats, *Biol. Neonate* **45:**125-128.

Gautier, C., and R. Vaillant, 1979, Postnatal changes in carbamoyl phosphate synthetase I and ornithine transcarbamylase activities after normal birth. Premature delivery and prolonged gestation in rat liver: Effects of actinomycin D and glucose, *Biol. Neonate* **35:**298-306.

Girard, J. R., P. Ferré, J. P. Pégorier, P. Turlan, L. El-Manoubi, and S. Callikan, 1981, Glucose metabolism in the newborn rat, *Biochem. Soc. Trans.* **9:**369-370.

Greengard, O., 1971, Enzymic differentiation in mammalian tissues, *Essays Biochem.* **7:**159-205.

Greengard, O., M. Federman, and W. E. Knox, 1972, Cytomorphometry of developing rat liver and its application to enzymic differentiation, *J. Cell Biol.* **52:**261-272.

Gregson, N. A., and P. L. Williams, 1969, A comparative study of brain and liver mitochondria from new-born and adult rats, *J. Neurochem.* **16:**617-626.

Hahn, P., and J. Skala, 1971, Development of enzyme systems, *Clin. Obstet. Gynecol.* **14:**655-668.

Hale, D. E., and J. R. Williamson, 1984, Developmental changes in the adenine nucleotide translocase in the guinea pig, *J. Biol. Chem.* **259:**8737-8742.

Hallman, M., 1971*a*, Changes in mitochondrial respiratory chain proteins during perinatal development. Evidence of the importance of environmental oxygen tension, *Biochim. Biophys. Acta* **253:**360-372.

Hallman, M., 1971*b*, Effect of intraperitoneal chloramphenicol on some mitochondrial enzymes in neonatal rats, *Biochem, Pharmacol.* **20:**1797-1809.

Hallman, M., and P. Kankare, 1971, Cardiolipin and cytochrome aa_3 in intact liver mitochondria of rats. Evidence of successive formation of inner membrane components, *Biochem. Biophys. Res. Commun.* **45:**1004-1010.

Hallman, M., and P. Kankare, 1979, Mitochondrial and microsomal phospholipid phosphorous metabolism during postnatal growth in rat heart and liver, *Lipids* **14:**435-440.

Hawkins, R. A., D. H. Williamson, and H. A. Krebs, 1971, Ketone-body utilization by adult and suckling rat brain *in vivo, Biochem. J.* **122:**13-18.

Hermann, H., and M. L. Tootle, 1964, Specific and general aspects of the development of enzymes and metabolic pathways, *Physiol. Rev.* **44:**289-371.

Herzfeld, A., M. Federman, and O. Greengard, 1973, Subcellular morphometric and biochemical analyses of developing rat hepatocytes, *J. Cell Biol.* **57:**475-483.

Hommes, F. A., 1975, Energetic aspects of late fetal and neonatal metabolism in *Normal and Pathological Development of Energy Metabolism,* F. A. Hommes and C. J. Van de Berg, eds., Academic Press, London, pp. 1-9.

Hommes, F. A., G. Luit-de Hann, and A. R. Richters, 1971, The development of some Krebs Cycle enzymes in rat liver mitochondria, *Biol. Neonate* **17:**15-23.

Hommes, F. A., and A. R. Richters, 1969, Mechanism of oxidation of cytoplasmic reduced nicotinamide adenine dinucleotides in the developing rat liver, *Biol. Neonate* **14:**359-364.

Hostetler, K. Y., B. D. Zenner, and H. P. Morris, 1979, Phospholipid content of mitochondrial and microsomal membranes from Morris hepatomas of varying growth rates, *Cancer Res.* **39:**2978-2983.

Holtzman, D., and C. L. Moore, 1973, Oxidative phosphorylation in immature rat brain mitochondria, *Biol. Neonate* **22:**230-242.

Jakovcic, S., J. Haddock, G. S. Getz, M. Rabinowitz, and H. Swift, 1971, Mitochondrial development in liver of foetal and newborn rats, *Biochem. J.* **121:**341-347.

Jones, C. T., ed., 1982, *The Biochemical Development of the Fetus and Neonate,* Elsevier Biomedical Press, Amsterdam.

Kelley, R. T., and J. R. Aprille, 1984, Postnatal accumulation of adenine nucleotides in rabbit liver mitochondria and the development of citrulline synthesis, *Pediatr. Res.* **18:**140A.

Kinnula, V. L., and I. Hassinen, 1977, Effect of hypoxia on mitochondrial mass and cytochrome concentrations in rat heart and liver during postnatal development, *Acta Physiol. Scand.* **99:**462-466.

Klingenberg, M., 1980, The ADP-ATP translocation in mitochondria, a membrane potential controlled transport, *J. Membrane Biol.* **56:**97-105.

Knowles, S. E., and F. J. Ballard, 1974, Pyruvate dehydrogenase activity in rat liver during development, *Biol. Neonate* **24:**41-48.

Krahling, J. B., R. Gee, J. A. Gauger, and N. E. Tolbert, 1979, Postnatal development of peroxisomal and mitochondrial enzymes in rat liver, *J. Cell Physiol.* **101:**375-390.

Krebs, H. A., R. Hems, P. Lund, D. Halliday, and W. W. C. Read, 1978, Sources of ammonia for mammalian urea synthesis, *Biochem. J.* **176:**733-737.

Lang, C. C., 1965, Respiratory enzymes in the heart and liver of the prenatal and postnatal rat, *Biochem. J.* **95:**365-371.

Lang, C. A., and G. H. Herbener, 1972, Quantitative comparison of the mitochondrial populations in the livers of newborn and weanling rats, *Dev. Biol.* **29:**176-182.

Levy, M., and R. Toury, 1970, Etude de L'evolution des activities des enzymes mitochondriaux de L'hepatocyte au cours du developpement du rat, *Biochim. Biophys. Acta* **216:**318-327.

Mäenpää, P. M., and N. C. R. Räihä, 1968, Content of cytochrome a + a_3 in the heart, liver, and brain of the developing rat, *Scand. J. Clin. Lab. Invest.* **21:**7.

Meister, R., J. Comte, L. Baggeto, C. Godinot, and D. C. Gautheron, 1983, Inhibitory effect of glucose on the maturation of rat liver mitochondria at birth. Phospholipid and oxidative metabolism, *Biochim. Biophys. Acta* **722:**36-42.

Mertens-Strijthagen, J., C. DeSchrijver, S. Wattiaux-DeConinck, and R. Wattiaux, 1979, A centrifugation study of rat liver mitochondria, lysosomes, and peroxisomes during the perinatal period, *Eur. J. Biochem.* **98**:339-352.

Miller. A. L., and P. Chu, 1970, The development of urea cycle enzyme activity in the liver of foetal and neonatal rats, *Enzym. Biol. Clin.* **11**:497-503.

Milner, R. D. G., and L. Pownall, 1981, Hepatic cellular development in the rabbit, *Biol. Neonate* **40**:204-208.

Mintz, H. A., D. H. Yawn, B. Safer, E. Bresnick, A. G. Liebelt, Z. R. Blailock, E. R. Rabin, and A. Schwartz, 1967, Morphological and biochemical studies of isolated mitochondria from fetal, neonatal, and adult liver and from neoplastic tissues, *J. Cell. Biol.* **34**:513-523.

Miyahara, M., Y. Kitazoe, N. Hiraoka, K. Takeda, S. Watanbe, J. Sasaki, E. Okimasu, Y. Osaki, H. Yamamoto, and K. Utsumi, 1984, Developmental changes in mitochondrial components in liver of newborn rats, *Biol. Neonate* **45**:129-141.

Murakami, A., Y. Kitagawa, and E. Sugimoto, 1983, In vivo study of developmental changes in carbamoyl phosphate synthetase I in rat liver. Repression of the enzyme synthesis immediately after birth, *Biochem. Biophys. Acta.* **739**:66-72.

Nakazawa, T., K. Asami, H. Suzuki, and O. Yukawa, 1973, Appearance of energy conservation system in rat liver mitochondria during development. The role of adenine nucleotide translocation, *J. Biochem.* **73**:397-406.

Ontko, J. A., 1966, Genesis of liver mitochondria in the neonatal rat, *Life Sci.* **5**:817-825.

Parameswaran, M., and I. J. Arinze, 1981, Relationship of mitochondrial phosphoenolpyruvate carboxykinase to neonatal gluconeogenesis, *Biol. Neonate* **39**:260-265.

Paul, H. S., and S. A. Adibi, 1982, Role of ATP in the regulation of branched-chain α-keto acid dehydrogenase activity in liver and muscle mitochondria of fed, fasted, and diabetic rats, *J. Biol. Chem.* **257**:4875-4881.

Pollak, J. K., 1975, The maturation of the inner membrane of foetal rat liver mitochondria. An example of a positive feedback mechanism, *Biochem. J.* **150**:477-488.

Pollak, J. K., 1976, Mitochondrial heterogeneity in foetal and suckling rat liver: Differential leucine incorporation into the proteins of two mitochondrial populations, *Biochem. Biophys. Res. Commun.* **69**:823-829.

Pollak, J. K., 1977, The interdependence of mitochondrial maturation and glycogen metabolism in perinatal rat liver, *Biochem. Soc. Trans.* **5**:341-348.

Pollak, J. K., and C. G. Duck-Chong, 1973, Changes in rat liver mitochondria and endoplasmic reticulum during development and differentiation, *Enzyme* **15**:139-160.

Prpic, V., and F. L. Bygrave, 1981, Maturation in liver mitochondria of Ruthenium Red-sensitive calcium-ion-transport activity and the influence of glucagon administration *in vivo* and *in utero*, *Biochem. J.* **196**:207-216.

Pollak, J. K., and E. A. Munn, 1970, The isolation by isopycnic density-gradient cerntrifugation of two mitochondrial populations from livers of embryonic and fed and starved adult rats, *Eur. J. Biochem.* **117**:913-919.

Pollak, J. K., and R. Sutton, 1978, The adenine nucleotide translocator in foetal, suckling and adult rat liver mitochondria, *Biochem. Biophys. Res. Commun.* **80**:193-198.

Pollak, J. K., and R. Sutton, 1980*a*, The differentiation of animal mitochondria during development, *Trends Biochem. Sci.* **5**:23-27.

Pollak, J. K., and R. Sutton, 1980*b*, The transport and accumulation of adenine nucleotides during mitochondrial biogenesis, *Biochem. J.* **192**:75-83.

Pollak, J. K., and R. Sutton, 1980*c*, The characterization of the carrier-system for ATP accumulation in rat liver mitochondria, *Austral. Biochem. Soc. Proc.* **13**:105.

Potter, V. R., W. C. Schneider, and G. J. Liebl, 1945, Enzyme changes during growth and differentiation in the tissues of the newborn rat, *Cancer Res.* **5**:21-24.

Rabier, D., P. Briand, F. X. Coude, F. Petit, G. Grimber, P. P. Kamoun, and L. Cathelineau, 1983, Developmental changes of citrullinogenesis, mitochondrial N-acetylglutamate content and N-acetylglutamate synthetase in fetal and neonatal rats, *Biochimie* **65:**1-6.

Räihä, N. C. R., and J. Suihkonen, 1968, Factors influencing the development of urea-synthesizing enzymes in rat liver, *Biochem. J.* **107:**793-797.

Rohr, H. P., A. Wirz, L. C. Henning, U. N. Riede, and L. Bianchi, 1971, Morphometric analysis of the rat liver cell in the perinatal period, *Lab. Invest.* **24:**128-139.

Rulfs, J., 1981, Regulation of fatty acid oxidation in neonatal rabbit liver mitochondria, Ph.D. thesis, Tufts University, Medford, Mass.

Rulfs, J., and J. R. Aprille, 1980, Regulation of short and medium chain fatty acid oxidation in newborn rabbit liver mitochondria, *Fed. Proc.* **39:**2114.

Rulfs, J., and J. R. Aprille, 1982, Adenine nucleotide pool size, adenine nucleotide translocase activity, and respiratory activity in newborn rabbit liver mitochondria, *Biochim. Biophys. Acta* **681:**300-304.

Sandor, K., and J. K. Pollak, 1976, The mechanism of mitochondrial maturation in the liver of the perinatal rat, *Biochem. Soc. Trans.* **4:**1122-1124.

Shelley, H. G., and G. A. Neligan, 1966, Neonatal hypoglycemia, *Brit. Med. Bull.* **22:**34-39.

Snell, K., 1974, Pyruvate carboxylation in liver mitochondria of the developing neonatal rat, *Internat. J. Biochem.* **5:**463-469.

Sutton, R., and J. K. Pollak, 1978, The increasing adenine nucleotide concentration and the maturation of rat liver mitochondria during neonatal development, *Differentiation* **12:**15-21.

Sutton, R., and J. K. Pollak, 1980, Hormone-initiated maturation of rat liver mitochondria after birth. *Biochem. J.* **186:**361-367.

Van Lelyveld, P. H., and F. A. Hommes, 1978, Adenine nucleotides in foetal rat liver cells. Compartmentation and variation with age, *Biochem. J.* **174:**527-533.

Van Rossum, G. D. V., 1963, Respiration and glycolysis in liver slices prepared from rats of different foetal and postnatal ages, *Biochim. Biophys. Acta* **74:**15-23.

Vergonet, G., F. A. Hommes, and I. Molenaar, 1970, A morphometric and biochemical study of fetal and adult rat liver cells, with special reference to energy metabolism, *Biol. Neonate* **16:**297-305.

Vernon, R. G., and D. G. Walker, 1968, Changes in activity of some enzymes involved in glucose utilization and formation in developing rat liver, *Biochem. J.* **106:**321-329.

Villee, C. A., and D. D. Hagerman, 1958, Effect of oxygen deprivation on the metabolism of fetal and adult tissues, *Am. J. Physiol.* **194:**457-468.

Watanabe, F., W. Kamiike, T. Nishimura, T. Hashimoto, and K. Tagawa, 1983, Decrease in mitochondrial levels of adenine nucleotides and concomitant mitochondrial dysfunction in ischemic rat liver, *J. Biochem.* **94:**493-499.

Wildman, S. G., T. Hongladarom, and S. I. Honda, 1962, Chloroplasts and mitochondria in living plant cells: Cinephotomicrographic studies, *Science* **138:**434-436.

Wilson, J. E., 1972, The relationship between glycolytic and mitochondrial enzymes in the developing rat brain, *J. Neurochem.* **19:**223-227.

4

Mitochondrial Regulation of Cytosolic Free Calcium

Elizabeth Murphy

National Institute of Environmental Health Sciences

INTRODUCTION

At present, the involvement of mitochondria in the physiological regulation of cytosolic free calcium (Ca_f) is quite controversial. A number of recent reviews have examined this topic (Rasmussen and Waisman, 1983; Borle, 1981; Denton and McCormack, 1980; Williamson, Cooper, and Hoek, 1981; Fiskum and Lehninger, 1982). In this review, we will evaluate the role of mitochondria in regulating Ca_f by the following criteria: (1) kinetics, (2) mitochondrial *set point* measurements (isolated and in situ mitochondria), (3) in situ total mitochondrial Ca contents, and (4) levels of Ca_f. The rationale employed is that if in situ mitochondria regulate Ca_f, their uptake and efflux pathways must be balanced at Ca_f. Thus by evaluating the interrelationship between levels of total mitochondrial Ca, Ca_f, and the level of extramitochondrial Ca to which the mitochondria regulate (the set point), we hope to determine the role of mitochondria in regulating Ca_f.

ROLE OF MITOCHONDRIA IN THE REGULATION OF BASAL CYTOSOLIC FREE CALCIUM

Kinetics

Since mitochondria were discovered to have a large Ca-accumulating capacity, investigators have frequently proposed mitochondrial involvement in regulating Ca_f (Vasington and Murphy, 1961; Lehninger, Carafoli, and Rossi, 1967). However, some recent studies have questioned the role of mitochondria in regulating basal levels of Ca_f (Brinley et al., 1977), given the low level of Ca_f coupled with the relatively high K_M for Ca uptake into mitochondria. At physiological concentrations of Mg, the K_M for Ca uptake into the mitochondria is reported to be in the range of 10 to 30 μM (Nicholls and Crompton, 1980; Carafoli and Crompton, 1978). Since several groups find Ca_f in a variety of tissues to be in the range of 0.03 to 0.2 μM (Brinley et al., 1977; Dipolo et al., 1976) it has been suggested that mitochondria may not be involved in the regulation of basal Ca_f.

Isolated Mitochondrial Set Point Measurement

Recent experiments by Nicholls (1978) and Becker, Fiskum, and Lehninger (1980) have elegantly demonstrated that mitochondria are able to precisely regulate extramitochondrial Ca and have shown that isolated mitochondria buffer extramitochondrial Ca to the set point (0.4 to 0.8 μM) (Becker, 1980). As seen in Figure 4-1, addition of Ca to a suspension of mitochondria (containing >10nmol Ca/mg mitochondrial protein) results in a rapid uptake of Ca until an extramitochondrial Ca level of 0.4 μM is achieved. Similarly, if the Ca concentration is lowered below 0.4 μM by the addition of ethyleneglycol-bis(β-aminoethylether)-N-N′-tetraacetic acid (EGTA), the mitochondria respond by releasing Ca until an extramitochondrial level of 0.4 μM is reached. Extramitochondrial Ca is maintained by coordination of the two mitochondrial Ca transport systems: the Ca uniporter which normally transports Ca into the mitochondria, and the Ca-H(Na) exchanger which normally extrudes Ca from the mitochondria. The uptake and release pathways are balanced at the set point (0.4 μM in Fig. 4-1). As the extramitochondrial Ca is raised, the uptake pathway is stimulated, while the efflux pathway rate remains essentially unchanged, resulting in Ca uptake into the mitochondria until Ca_f falls below a level which stimulates uptake (the set point is reached). As the extramitochondrial Ca is lowered below the set point, the efflux pathway releases at essentially the same rate but Ca uptake via the uniporter is now depressed or stopped, resulting in a net Ca efflux.

It is clear that for mitochondria to maintain extramitochondrial Ca at a constant value, it is essential that the rate of the Ca efflux pathway remains constant as the mitochondrial Ca content changes due to net Ca uptake or release.

Two conditions under which the efflux pathway could be constant when the total mitochondrial Ca is changing are: (1) if the efflux pathway is operating at saturation (therefore increasing or decreasing intramitochondrial

Ca would not alter the efflux rate), or (2) if the level of intramitochondrial free Ca does not change when there is a change in the level of mitochondrial Ca (i.e., the mitochondria buffer changes in Ca so that altering total Ca would have little effect on free Ca). However, several reports (Williamson and Murphy, 1980; Hansford and Castro, 1982; Coll et al., 1982) suggest that as total mitochondrial Ca increases, free intramitochondrial Ca also increases, making the latter possibility unlikely. The present consensus is that mitochondria buffer extramitochondrial Ca only when the efflux pathway is saturated. An examination of the relationship between total mitochondrial Ca and extramitochondrial free Ca suggests that rat liver mitochondria with >4 nmol Ca/mg mitochondrial protein can effectively buffer extramitochondrial Ca (see Hansford and Castro, 1982; Joseph et al., 1983; and Fig. 4-2). As shown in Figure 4-2 (also see Joseph et al., 1983), isolated mitochondria which contain more than 4 nmol Ca/mg mitochondrial protein are capable

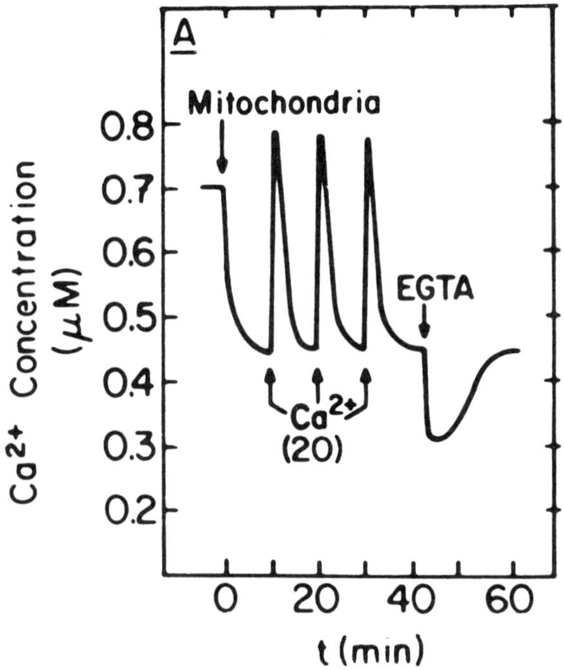

Figure 4-1. Transport-mediated regulation of the steady state ambient free Ca^{2+} concentration by rat liver mitochondria. Rat liver mitochondria (5.0 mg of protein) were added to the test medium (5.0 ml total volume, 25°C). The number in parentheses is the addition of Ca^{2+} in nanogram-ions/mg of mitochondrial protein. The level of added EGTA was 10 nmol/mg of protein (from Becker et al., 1980).

of maintaining a more or less constant set point. Mitochondria containing less than 4 nmol Ca/mg mitochondrial protein are in equilibrium with Ca_f (i.e., their influx and efflux pathways are balanced); however, alterations in Ca_f would cause the mitochondria to take up or release Ca, thus altering the efflux pathway rate which would alter their set point. In other words, under these conditions rat liver mitochondria containing less than 4 nmol Ca/mg protein (unsaturated efflux pathway) are in equilibrium with Ca_f but will not maintain Ca_f at a constant level.

It is of interest to determine whether mitochondria in situ maintain Ca_f at a constant level (i.e., whether the efflux pathway is saturated by mitochondrial Ca greater than 4 nmol Ca/mg protein), or whether the level of mitochondrial Ca is a function of Ca_f. To address this question, we will (1) examine studies that report the in situ level of mitochondrial Ca to determine whether it is above 4 nmol Ca/mg protein, (2) examine the ability of in situ mitochondria to regulate Ca_f, and (3) examine levels of Ca_f to determine if they are compatible with the mitochondrial efflux pathway operating at saturation.

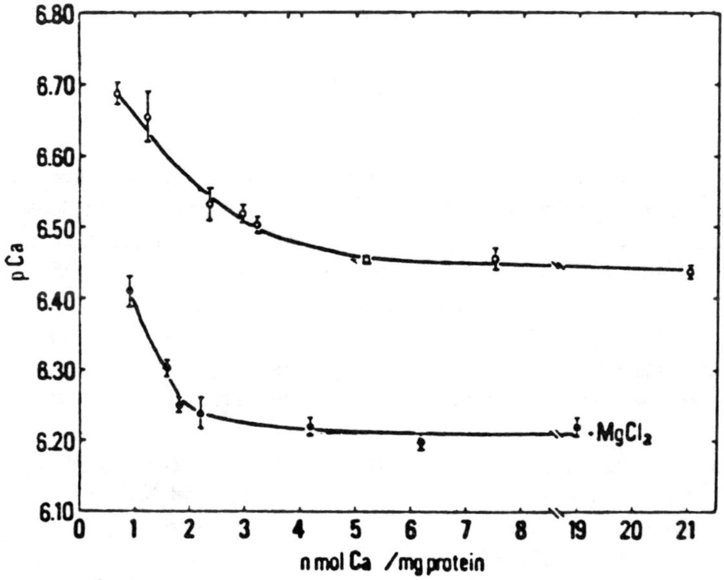

Figure 4-2. Dependence of extramitochondrial pCa on the Ca content of heart mitochondria. Results are mean values ± SEM from experiments with five mitochondrial preparations. Mitochondria (6.25 mg protein) were added to 4.1 ml of medium. The temperature was 25°C and the pH was 7.2. Where indicated, $MgCl_2$ was 1 mM (from Hansford and Castro, 1982).

Level of Mitochondrial Calcium In Situ

The in situ level of mitochondrial Ca is key to defining the role of mitochondria in regulating Ca_f. Different approaches addressing the in situ level of mitochondrial Ca are discussed below.

Regulation of Intramitochondrial Enzymes. Denton and co-workers (Denton and McCormack, 1980; McCormack and Denton, 1980; Denton, McCormack, and Edgell, 1980) and Hansford and Castro (1982) have estimated in situ mitochondrial Ca levels based on the Ca sensitivity of several mitochondrial enzymes: 2-oxyglutarate dehydrogenase, pyruvate dehydrogenase, and isocitrate dehydrogenase. McCormick and Denton (1980) used uncoupled rat heart mitochondria in the presence of the Ca ionophore A23187 and assumed that the extramitochondrial Ca was equal to the intramitochondrial Ca. The extracellular Ca was then varied using Ca-EGTA buffers in the presence of Mg-ATP and oligomycin. By extracting the pyruvate dehydrogenase from these mitochondria and determining the proportion in the active form, they determined a $K_{0.5}$ of 1 μM for Ca activation of pyruvate dehydrogenase. This suggests that unless pyruvate dehydrogenase normally functions at maximal activity, the intramitochondrial free Ca must at least occasionally be less than 1 μM.

Measurements of intramitochondrial free Ca have been made using a null point technique (Williamson and Murphy, 1980; Hansford and Castro, 1982; Coll et al., 1982) and there appears to be a linear relationship between total and free mitochondrial Ca. The mitochondrial *null point* measures the extramitochondrial free Ca level at which there is no uptake or release of Ca upon addition of the Ca ionophore, A23187. A free mitochondrial Ca level of 1 μM is compatible with a total mitochondrial Ca content in the range of 0.25 to 1.0 nmol/mg mitochondrial protein (Values differ depending on correction for any residual ΔpH after A23187 addition; see Coll et al., 1982). More importantly, the 1 μM free Ca level and the total Ca with which it is compatible appear to be below that required to saturate the efflux pathway (see Fig. 4-2).

In another set of experiments using intact, energy-coupled rat heart mitochondria in the presence of physiological Mg and Na concentrations, Denton, McCormack, and Edgell (1980) reported that an extramitochondrial free Ca level of 0.5 μM gave maximal activation of pyruvate dehydrogenase. Thus at Ca_f levels below 0.5 μM (where most values for Ca_f are) pyruvate dehydrogenase activity is not saturated.

Hansford and Castro (1982) have reported that 2-oxyglutarate dehydrogenase is activated over the range of 0 to 2 nmol Ca/mg mitochondrial protein (or ~0-1.2 μM free Ca as measured by the mitochondrial null point technique). The assumption that mitochondria in situ operate in a range of Ca that does not saturate the dehydrogenase suggests an in situ total mitochondrial Ca level of less than 2 nmol/mg protein. At this level of Ca, the efflux pathway is not saturated and, as seen in Figure 4-2, the set point has still not plateaued.

In summary, the data obtained with Ca-dependent intramitochondrial enzymes suggest that their activities are saturated at mitochondrial Ca levels compatible with mitochondrial maintenance of a constant Ca_f.

Uncoupler-releasable Cellular Calcium and Fractionation Techniques. Mitochondrial uncouplers collapse the mitochondrial membrane potential, resulting in Ca release. Addition of the uncoupler FCCP to intact cells causes release of Ca from the cells. FCCP-releasable Ca has been used by many groups (Joseph et al., 1983; Murphy et al., 1980; Blaustein, Ratzlaff, and Schweitzer, 1978; Murphy and Mandel, 1982; Scott, Akerman, and Nicholls, 1980; Mandel and Murphy, 1984) as an estimate of in situ mitochondrial Ca. Subsequent measurement of A23187-releasable Ca has been used as an index of microsomal Ca. The estimates of in situ organelle Ca by FCCP and A23187 release have compared favorably with Ca measurements of mitochondria and microsomes isolated by rapid cell fractionation (Murphy et al., 1980; Scott, Akerman, and Nicholls, 1980). Murphy and co-workers (1980) reported the FCCP-releasable Ca in isolated hepatocytes (containing a total Ca of 3-5 nmol/mg cell protein) to be 60 to 70% of total cell Ca. Approximately 20% of the total cell Ca of hepatocytes was microsomal. Liver cells were also rapidly fractionated by treatment with digitonin and shearing force in the presence of ruthenium red and EGTA, followed by centrifugation through silicone oil. This resulted in a pellet containing mitochondria as well as membrane fragments and cell debris. The Ca content of this pellet was in good agreement with the FCCP-releasable Ca from liver cells. Assuming 25% of the total cell protein to be mitochondrial, this suggests that in situ mitochondria contain about 10 nmol Ca/mg mitochondrial protein (65% $\times$ 4 nmol/mg $\div$ 25% = 10.4 nmol/mg mitochondrial protein).

Joseph et al. (1983) took this general approach one step further by measuring the cytosolic free Ca concentration of rat hepatocytes and correlating that measurement with the intracellular Ca distribution. The cytosolic Ca_f was measured by a null point technique similar to that previously described for measuring the mitochondrial matrix Ca_f. The total Ca content of the hepatocytes was lowered by incubation with EGTA and elevated by incubation with high levels of extracellular Ca in the cold. They found an increase in Ca_f from 0.1 to 0.3 μM as they increased total cell Ca from 1 to 5 nmol/mg dry wt. Above 5 nmol/mg dry wt the Ca_f remained at about 0.3 μM. Using FCCP and A23187 to estimate the mitochondrial and microsomal Ca levels, they reported that as total Ca increased from 1 to 5 nmol/mg dry wt, both the mitochondrial and microsomal pools increased. As total cell Ca increased above 5 nmol/mg dry wt, the mitochondrial Ca increased but the microsomal Ca showed no further increase. They reported basal levels of total cell Ca to be 5 to 6 nmol/mg dry wt with about 4 nmol/mg dry wt releasable by FCCP, or about 16 nmol Ca/mg mitochondrial protein (assuming 0.25 mg mitochondrial protein/mg cell dry wt). They concluded from these data that at the basal level of total cell Ca the microsomes are saturated with Ca and the mitochondria contain sufficient Ca to be an effective Ca buffer system (see Fig. 4-2).

Murphy and Mandel (1982), using uncoupler-releasable Ca, showed that proximal kidney tubules contain about 80% of the total basal Ca content in the mitochondria. Basal Ca levels in these cells were approximately 13 to 14 nmol Ca/mg protein, with 10 to 12 nmol Ca/mg cell protein released by FCCP. Assuming that 25% of the total cell protein is mitochondrial, these numbers translate into 40 nmol total cell Ca/mg mitochondrial protein. This level of mitochondrial Ca is well within the range that would allow mitochondrial buffering of Ca_f. Measurement of Ca_f for these tubules gave a basal level of 0.45 μM. More recent studies by Mandel and Murphy (1984) report that about 45 to 60% of the total cell Ca ($\sim$10–17 nmol total cell Ca/mg cell protein) is mitochondrial, or about 20 to 40 nmol/mg mitochondrial protein (an amount sufficient to buffer Ca_f).

It has been suggested that FCCP will reduce cell ATP, thereby reducing microsomal Ca and resulting in an overestimation of mitochondrial Ca (Burgess et al., 1983). To address this concern, Mandel and Murphy (1984) examined ruthenium red-induced Ca release in rabbit proximal kidney tubules. Although ruthenium red is usually impermeant to cells (Luft, 1971), it appears to enter proximal kidney tubules (Mandel and Murphy, 1984; McCormick, and England, 1983). The intercellular action of ruthenium red appears to be specific for mitochondria since Bygrave (1978) showed it had no effect on Ca accumulation by endoplasmic reticulum. Ruthenium red caused 30% of the cellular Ca to be released from the tubules in one hour while paired controls retained their Ca level over this time period (30% of 10 nmol/mg ÷ 25% mitochondrial protein = 12 nmol/mg mitochondrial protein). Ruthenium red also markedly inhibited the rate of Ca release by FCCP, thus substantiating mitochondria as the primary intracellular Ca store. Ruthenium red does not release microsomal Ca (Bygrave, 1978), suggesting that in situ kidney mitochondria contain at least 12 nmol Ca/mg mitochondrial protein. This is sufficient Ca for the mitochondria to buffer Ca_f.

Scott and co-workers (1980) and Akerman and Nicholls (1981, 1983) determined the distribution of Ca in synaptosomal mitochondria and microsomes using a rapid fractionation technique similar to that described earlier

(Murphy et al., 1980). Ruthenium red and EGTA were included in the disruption solution and the synaptosomes were centrifuged through silicone oil within 30 seconds. As mentioned earlier, the resultant pellet contained cellular contaminants in addition to mitochondria. In control experiments, the ^{45}Ca content was measured in the presence of rotenone and oligomycin (to block mitochondrial Ca uptake) and in the presence of oligomycin alone (to allow respiration-dependent mitochondrial Ca uptake but to inhibit ATP production, thereby excluding microsomal ^{45}Ca uptake). Scott, Akerman, and Nicholls (1980) reported control synaptosomal Ca contents of around 13.8 nmol Ca/mg protein; the pellet of disrupted control cells contained about 10.8 nmol Ca/mg protein, or about 79% of the total Ca. Treating synaptosomes with rotenone and oligomycin reduced the pellet after disruption to 2.4 nmol Ca/mg protein, suggesting that about 75 to 80% of the Ca in the pellet is mitochondrial. In the presence of oligomycin alone, total synaptosomal Ca was 16.7 nmol Ca/mg protein with 83% in the disrupted pellet, consistent with little contribution of microsomal Ca to the Ca in the pellet after disruption.

Williamson and co-workers (1983), using freshly disaggregated (Ca-tolerant) adult rat heart cells, reported that FCCP released ⅓ of the total releasable Ca. FCCP released 1.2 nmol Ca/mg dry wt from cells containing 4.6 nmol Ca/mg cell dry wt or 26% of the total cell Ca. The 1.2 nmol Ca/mg dry wt released translates into 4.8 nmol Ca/mg mitochondrial protein, using a value of 0.25 mg mitochondrial protein /mg cell dry wt. The authors suggest that this level of mitochondrial Ca is sufficient to buffer Ca_f.

Using cultures of embryonic chick heart cells, Murphy and co-workers (1983) determined FCCP releasable Ca to be only 0.8 ± 0.9 nmol Ca/mg cell protein out of a total cell Ca of 4.2 ± 0.7 nmol Ca/mg cell protein. This finding suggests that the basal mitochondrial Ca level is considerably less in embryonic heart than in liver (Joseph et al., 1983; Murphy et al., 1980) or kidney (Murphy and Mandel, 1982; Mandel and Murphy, 1984). The Ca level would be $0.8 \times 4 = 3.2$ nmol/mg mitochondrial protein, but given the significant variability in the results, it is difficult to conclude that this level of mitochondrial Ca is sufficient to buffer Ca_f. However, incubating the cells with ouabain for one hour resulted in an increase in the total cell Ca to 35.0 ± 4.6 nmol Ca/mg cell protein, concomitant with a rise in mitochondrial Ca to 15.8 ± 4.7 nmol Ca/mg cell protein. These results suggest that under Ca-loading conditions heart mitochondria do increase their Ca content in situ. In other studies, Murphy, Jacob, and Lieberman (1984) showed that addition of ouabain for 10 minutes elevates Ca_f by 2.5- to 4-fold, as measured by either the fluorescent indicator Quin2 or the null point technique.

Brinley and co-workers (1977) examined the intracellular Ca buffering in squid axon by measuring absorbance changes of the Ca-sensitive dye arsenazo

III, which had been pressure-injected into the axon. For calibration, arsenazo III was placed in a quartz glass capillary with a diameter similar to that of the axons (600 μm). Ca-EGTA buffers present in an ionic environment similar to that present within the squid axon were used to calibrate the absorbance changes. Axons loaded with Ca by various manipulations (e.g., micro-injection, stimulation in 112 mM Ca sea water, or soaking in an Na-free solution with 1-10 mM Ca) were able to effectively buffer 99.9% of the imposed Ca loaded (the ionized Ca rose only 0.6 nM/μM load). When similar Ca loads were imposed on FCCP-treated axons, 93 to 95% of the Ca loaded was buffered. When FCCP was added to a fresh intact axon containing only endogenous Ca (50 μmol Ca/kg axoplasm), Ca_f rose from 30 nM to 170 nM. Thus, addition of FCCP resulted in a net release of 140 nM. Since in the presence of FCCP, 95% of an imposed Ca loaded is buffered and does not appear as an increase in Ca_f, the value of 140 nM was multiplied by 20 to obtain the endogenous mitochondrial Ca value of 2.8 μmol/kg axoplasm. Using a value of 1% for the mitochondrial volume of the axoplasm, a value of 0.28 mmol Ca/kg mitochondria wet wt can be calculated. Using the conversion factor that 1 kg mitochondrial protein = 3.43 kg wet wt (dry mass mitochondria = 70% protein, 1 kg dry wt mitochondria = 2.4 kg wet wt), they concluded that in situ mitochondria have ~1 mmol Ca/kg mitochondrial protein. A mitochondrial Ca level of 1 mmol/kg protein (1 nmol/mg) would be too low to buffer Ca_f. In these experiments in which FCCP addition was made to axons in a Ca-free solution, Ca may be released from the cells, thus underestimating the mitochondrial Ca. Brinley and co-workers (1977) point out that even if the total axoplasm Ca (50 μmol/kg axoplasm) were contained in the axon mitochondria, the mitochondrial Ca level would still only be 17 mmol/kg protein. It is worth noting the low total cell Ca of the axon (50 μmol/kg) compared to tissues such as heart (Williamson et al., 1983; Murphy et al., 1983; Shen and Jennings, 1972), kidney (Murphy and Mandel, 1982; Mandel and Murphy, 1984; Borle and Clark, 1981), and liver (Joseph et al., 1983; Murphy et al., 1980), where the total cell Ca ranges from 3 to 10 mmol/kg. This is particularly interesting given the 25- to 30 fold difference between the squid axon and the tissues specified in mitochondrial volume as a percentage of total cell volume.

Brinley and co-workers (1977) also loaded axons with Ca by incubation for 12 minutes in a solution containing 1 mM Ca and no Na. The accumulated Ca load was calculated to be 48 μmol/kg axoplasm. Subsequent addition of FCCP (again with no Ca_o) caused release into the cytoplasm of 850 nM Ca. Again correcting for the 95% Ca buffering in the presence of FCCP, they calculated that 17 μmol Ca/kg axoplasm were contained in the mitochondria. Subtracting the endogenous mitochondrial Ca of ~3 μmol Ca/kg axoplasm gave 14 μmol Ca/kg axoplasm of the imposed load sequestered by the mitochondria. If the 48 μmol Ca/kg axoplasm estimate of the total Ca load is accepted, probably 14/48 or 29% of the load is sequestered by the mitochondria.

Most of the data obtained either by rapid cell fractionation techniques (Murphy et al., 1980; Scott, Akerman, and Nicholls, 1980) or from FCCP-releasable Ca measurements (Joseph et al., 1983; Murphy et al., 1980; Murphy and Mandel, 1982; Mandel and Murphy, 1984; Williamson et al., 1983) indicate that the level of mitochondrial Ca in situ is greater than 4 nmol/mg mitochondrial protein. As shown earlier, these levels of total mitochondrial Ca are compatible with a mitochondrial set point in the range to buffer Ca_f. There are other data, however, suggesting that in situ mitochondrial Ca is lower. The data of Murphy and co-workers (1983), measuring the FCCP-releasable Ca in embryonic chick heart, show a Ca content marginal for mitochondrial buffering of Ca_f. The data of Brinley and co-workers (1977) suggest a total mitochondrial Ca content in the squid axon of ~1 nmol/mg mitochondrial protein, a level too low for mitochondrial buffering of Ca_f. However, in the experiments of Brinley and co-workers (1977), the FCCP-releasable Ca was measured intracellularly; all other experiments measure Ca release from the cells. Since the experiments of the Brinley group were done using nominally Ca_o-free solutions, they could have underestimated the FCCP-releasable Ca due to Ca efflux from the cell. However, they did consider the fact that in the presence of FCCP, only 5% of the Ca load appears as an increase in Ca_f, which presumably corrects for any Ca released by FCCP not appearing in the cytosol.

In conclusion, while most of the data are consistent with a role for mitochondria in buffering Ca_f, the squid axon data suggest that mitochondrial Ca is too low to buffer Ca_f. Whether these discrepancies result from methodologic differences or tissue differences remains to be determined.

Electron Probe Measurements. Somlyo and co-workers, using electron-probe analysis, have suggested that in situ mitochondria contain low levels of Ca (Somlyo, Shuman, and Somlyo, 1981; Kracke et al., 1980; Somlyo et al., 1981; Somlyo, Somlyo, and Shuman, 1979), and furthermore, that mitochondria do not appear to accumulate Ca even in tetanized muscle (Somlyo et al., 1981) or during sustained K contracture (Somlyo, Somlyo, and Shuman, 1979). Quick-frozen ultrathin cryosections of the portal-anterior mesenteric vein (PAMV) were probed for the presence of mitochondrial Ca (Somlyo, Somlyo, and Shuman, 1979). Control mitochondrial Ca, which was not significantly different from total Ca in the cytoplasm, was 0.8 ± 0.5 mmol/kg dry wt (mean $\pm$ SE). Even after 30 minutes of high K contracture (maintaining a tonic component 80% of peak contraction), no increase in mitochondrial Ca as compared to the cytoplasm was found. However, both cytoplasmic and mitochondrial Ca were elevated. Similarly, Somlyo and co-workers (1981) using quick-frozen ultrathin sections of paired fiber bundles from frog (*Rana pipiens*) reported no difference in mitochondrial Ca during rest and after a 1.2s tetanus (mitochondrial Ca means ± SD in mmol/kg dry wt, rest, 1.7 ± 3.5, after tetanus, 0.5 ± 3.7). The paried values for Ca in the cytoplasm were: resting 3.0 ± 5.8 and tetanus, 7.2 ± 5.0.

In other studies, Somlyo and co-workers (1978) using electron probe analysis on isolated rabbit PAMV mitochondria found three groups of mitochondria: some with no granules, some with hard granules, and some with fluffy granules. The mitochondria with no granules contained 4.9 ± 0.7 mmol Ca/kg dry wt (mean ± SE), those with hard granules had 704.8 ± 4.8 mmol Ca/kg dry wt, and those with fluffy granules had 210.7 ± 3.1 mmol Ca/kg dry wt. They further reported that isolated mitochondria without granules had a calcium content similar to that found in situ. A Ca content of 4.9 nmol/mg mitochondrial dry wt has sufficient Ca to buffer Ca_f. Somlyo, Bond, and Somlyo (1985) have recently reported electron probe microanalysis measurement of in situ mitochondrial calcium levels in isolated hepatocytes. They report total mitochondrial calcium levels of 0.8 ± 0.1 mmol per mg mitochondrial dry wt, which is equivalent to 1.1 nmol per mg mitochondrial protein, a level below that required to saturate the mitochondrial efflux pathway.

Summary. There is considerable disagreement over the in situ mitochondrial Ca level. Assuming the mitochondrial dehydrogenases are not saturated in vivo, these data suggest a mitochondrial Ca content of less than 2 nmol/mg mitochondrial protein. Electron probe analysis data suggest that in situ basal mitochondrial Ca levels are less than about 3 to 8 nmol Ca/mg mitochondrial protein. With the exception of the squid axon data, most of the data for in situ mitochondrial Ca content obtained by either rapid fractionation or FCCP release suggest mitochondrial Ca levels (10-40 nmol/mg mitochondrial protein) sufficient to be involved in buffering Ca_f.

How can this disparity in the data be reconciled? In general, all of these measurements have potential errors that are not insignificant. Therefore it is difficult, based on the available data, to accept or reject a mitochondrial Ca content of 4 nmol/mg mitochondrial protein. Possible reasons for such order-of-magnitude differences include:

1. FCCP reduces ATP and may cause a release of microsomal Ca, thus causing overestimates of mitochondrial Ca levels. However, Mandel and Murphy (1984) demonstrated a similar release of Ca from cells using ruthenium red. Also, fractionation studies support the relatively high mitochondrial Ca level.
2. The isolated cells on which the measurements of in situ Ca were made

may have accumulated Ca during isolation. LeFurgey et al. (in press) have recently found by electron probe x-ray microanalysis that the vast majority of isolated proximal tubules have a normal intracellular K/Na of ~4 but contain no detectable mitochondrial calcium. However, in this same preparation ~5% of the tubules contain an abnormally low intracellular K/Na of ~1, and have large levels of calcium in the mitochondria. This high level of calcium in these damaged cells would artifactually elevate the apparent in situ level of mitochondrial calcium measured in a bulk suspension of these tubules.

3. Differences in ionic environment may alter Ca homeostasis (e.g., Mg, ATP, ionic strength, pH, etc.).
4. Tissue-specific differences may be involved.
5. EGTA-Ca binding constants may vary.

In conclusion, the squid axon data suggest total mitochondrial Ca levels of ~1 nmol/mg mitochondrial protein; the estimates based on mitochondrial dehydrogenase are also in the range 1 to 2 nmol/mg protein. Data for other tissues suggest higher levels of in situ mitochondrial Ca. However, until these differences in the level of mitochondrial Ca are resolved, it will be impossible to use this criterion to establish a role for or against mitochondrial regulation of Ca_f.

Calcium Uptake by In Situ Mitochondria and Microsomes

Becker, Fiskum, and Lehninger (1980) showed that isolated mitochondria have a set point of ~0.45 μM (see Fig. 4-1). Addition of isolated rat liver microsomes to isolated mitochondria lowers the set point to 0.2 μM Ca. If the isolated mitochondria are first treated with EGTA to lower their total Ca, the isolated mitochondria then maintain Ca at a set point of 0.2 μM. Becker and co-workers also showed that a mixture of microsomes and EGTA-treated mitochondria maintain a set point of 0.2 μM Ca but that after several additions of Ca a rise in the set point occurs until the Ca capacity of the microsomes is exceeded and then the system buffers at 0.4 μM Ca.

Mandel and Murphy (1984) examined the ability of intracellular organelles to regulate Ca_f by permeabilizing kidney proximal tubules with digitonin in a nominally Ca-free solution. As in the experiments of Becker, Fiskum, and Lehninger (1980), digitonin caused the equilibration of intracellular and extracellular Ca, allowing the intracellular Ca buffers to reach a set point prior to FCCP addition. This set point is approximately 0.5 μM in a NaCl or KCl medium. FCCP was then added to determine the total mitochondrial Ca content, which in this case was 45 nmol/ml (about 10-15 nmol/mg cell protein). After the initial set point was reached, a total of 50 nmol Ca/ml was added to the external medium. All of the added Ca was sequestered by the intracellular buffers and the measured free Ca returned to the same set point. Upon addition of FCCP, 95 nmol Ca/ml were released, constituting the original amount plus 100% of the accumulated Ca, suggesting that the Ca had been sequestered by the mitochondria. Similarly, after addition of digitonin, a total of 20 nmol/ml of EGTA was added, causing the release of Ca from the intracelluar buffers in an attempt to maintain the set point at about 0.5 μM. Addition of FCCP caused the release of only 30 nmol Ca/ml, indicating that the Ca released to maintain the set point had originated from the mitochondria. Therefore, if the free Ca concentration is raised, the mitochondria accumulate Ca to lower the extramitochondrial Ca. Likewise, if the free Ca is lowered, it is again the mitochondria that release Ca to raise the extramitochondrial Ca.

Streb and Schultz (1983) reported that in the presence of Mg-ATP and oxidizable substrates, addition of leaky acinar pancreas cells reduced the Ca concentration of a low Ca medium to 0.42 ± 0.01 μM, as measured by a Ca ion-selective electrode. Addition of mitochondrial inhibitors (10^{-5} antimycin, 10^{-2} azide and 5×10^{-6} M oligomycin) reduced the rate of uptake but the final set point was unaltered. Addition of vanadate (an inhibitor of Ca ATPase) at a concentration ineffective on mitochondrial Ca uptake raised the Ca set point to 0.68 μM. The authors interpreted these data as suggesting that the Ca_f level is maintained by a nonmitochondrial pool.

Burgess and co-workers (1983) came to a similar conclusion using guinea pig hepatocytes treated with saponin to measure the effect of mitochondrial and microsomal inhibitors on ^{45}Ca uptake. With 20 nM extracellular ^{45}Ca and an ATP regenerating system present, ^{45}Ca uptake was the same in the presence or absence of DNP and oligomycin. Similarly, with 20 nM ^{45}Ca and constant ATP, the ^{45}Ca uptake was identical with and without caffeine. The interpretation of the Burgess group was that the mitochondria cannot take up Ca at low extramitochondrial Ca levels, a conclusion reached by defining mitochondrial uptake as ^{45}Ca uptake in the absence of DNP + oligomycin minus ^{45}Ca uptake in the presence of DNP + oligomycin. However, since ^{45}Ca uptake was the same with and without caffeine, either caffeine is not effective at inhibiting endoplasmic reticulum (ER) Ca accumulation in rat liver, or organelles other than ER have Ca exchange mechanisms active at 20 nM Ca.

Studies by Brinley, Tieffert, and Scarpa (1978) examined the ability of squid axon mitochondria in situ to buffer Ca_f. The axons were loaded with Ca sensitive dyes and calibrated as already described. Axons were loaded with Ca to a level of 1 μM (by micro-injection, stimulation, or incubation in a

Na-free medium). The bathing solution was then changed to one that contained neither Ca nor Na in order to minimize the effects of plasma membrane Ca transport systems on the level of cytoplasmic Ca (as measured by arsenazo III). They demonstrated essentially no mitochondrial Ca uptake at Ca_f levels below 300 nM over a period of 10 min. However, as Ca_f increased above 300 nM, Ca uptake by the presumed mitochondrial component increased rapidly.

The threshold level of ~300 nM reported by Brinley, Tieffert, and Scarpa (1978) for Ca uptake by mitochondria is consistent with the mitochondrial set point data showing that isolated mitochondria buffer Ca at ~0.4 μM (Becker, Fiskum, and Lehninger, 1980). However, the squid axon data show that basal levels of Ca_f are around 30 to 50 nM. If Ca_f is maintained at this level it is unlikely that mitochondria are involved in buffering Ca_f. Moreover, at levels of 30 to 50 nM, mitochondria should release Ca (Fig. 4-1). Figure 4-2 shows an extramitochondrial Ca level of 50 nM compatible with a total mitochondrial Ca content of less than 0.5 nmol/mg protein, a value less than 50% of the estimated mitochondrial Ca content of squid axon.

In summary, the data in this section are conflicting as to whether or not in situ mitochondrial are involved in buffering Ca_f. The data of Mandel and Murphy (1984) support this idea but on the other hand the study of Streb and Schultz (1983) shows that inhibition of microsomal Ca uptake elevates the Ca set point. This finding suggests that mitochondria do not contribute to basal regulation of Ca_f. Clearly, further experimentation is needed to resolve these differences.

Measurements of Ca_f

As already discussed, the data with isolated mitochondria (or digitonin-treated cells) suggest that mitochondria effectively buffer Ca_f at approximately 0.4 μM; the exact value varies from study to study, depending on conditions, and falls within the range of 0.3 to 0.8 μM. Therefore, a situation in which Ca_f is maintained substantially below 0.4 μM strongly argues against a role for mitochondria in buffering basal Ca_f. A sampling of data concerning levels of Ca_f as measured by a variety of techniques follows.

Intracellular Dyes. As already discussed, experiments measuring Ca_f in squid axons suggest a basal level of 30 to 50 nM (Brinley et al., 1977; Brinley, Tieffert, and Scarpa, 1978; Requena et al., 1977). These measurements were determined using the metallochromic dyes arsenazo III and antipyrylazo III, and aequorin in a porous dialysis capillary inserted through the axon.

Ca-selective Microelectrodes. The following measurements were made using Ca-microelectrodes: Marban and co-workers (1980) measured a mean

Ca_f of 260 nM in quiescent ventricular (ferret papillary) muscle; Sheu and Fozzard (1982) reported a Ca_f of 270 nM in sheep ventricular muscle; Lee, Taylor, and Windhager (1980) reported a Ca_f of 120 nM in *Necturus* kidney; and O'Doherty and Stark (1982) found a Ca_f of 440 nM in parotid salivary gland.

Quin2. The intracellular fluorescent dye Quin2 has been used in a number of recent studies measuring Ca_f. Ca_f levels obtained using Quin2 are as follows: Tsien, Possan, and Rink (1982), 120 nM basal level in lymphocytes; Possan and co-workers (1983), 126 nM mean resting value in neutrophilis; Charest and co-workers (1983), 200 nM in unstimulated liver cells; Hesketh and co-workers (1983), 110 nM basal level in lymphocytes; Wollheim and Possan (1984), 105 nM mean resting Ca in an insulin-secreting cell line; Murphy, Jacob, and Lieberman (1984), 70 nM in freshly disaggregated embryonic chick heart cells.

Null Point Method. The following values were obtained using a null point technique. Murphy and co-workers (1980) reported a Ca_f of 180 nM in freshly disaggregated liver cells. Simons (1981) found 180 nM Ca_f in human red blood cells and 280 nM Ca_f in pigeon red cells. Streb and Schultz (1983) determined a Ca_f in rat acinar pancreas of 420 nM. Murphy and Mandel (1982) reported a Ca_f of 450 nM in isolated proximal kidney tubules. Murphy et al. (unpublished) measured a Ca_f of 230 nM in embryonic heart cells and Williamson and co-workers (1983) reported a Ca_f of 270 nM in Ca tolerant adult myocytes.

HOST Method. Borle and Snowdowne (1982) described a method of hypo-osmotic shock for introducing aequorin into cells. Using this method a value of 57 nM was reported for Ca_f in cultured monkey kidney cells.

Summary. It is obvious from the preceding material that there is a wide range of values for Ca_f (30–450 nM). Levels of Ca_f in the range of 30 to 100 nM, measured by Quin2 or by dyes injected into squid axon, are too low to be compatible with mitochondrial buffering of Ca_f (see Fig. 4-2). However, the higher levels of Ca_f measured by both Ca microelectrodes and the null point are compatible with a role for mitochondria in buffering Ca_f. However, until the discrepancies among these methods are resolved it will be difficult to determine on the basis of the Ca_f level the role of mitochondria in regulating Ca_f.

CONCLUSIONS

The purpose of this review has been to examine the role of mitochondria in the regulation of Ca_f and to investigate in particular the data for and against the possibility that in situ mitochondria operate with a saturated efflux pathway to maintain a constant Ca_f. This point has important ramifications for determining how cells elevate Ca_f during hormonal stimulation. If Ca_f is set by the plasma membrane transporters with mitochondria in equilibrium with (but not buffering) Ca_f (i.e., mitochondria operating with a nonsaturated efflux pathway), then an increased Ca influx across the plasma membrane would result in an increase in mitochondrial Ca with an increase also in the set point until the mitochondrial efflux pathway is saturated. Under basal conditions, the mitochondria would be coordinated with other Ca regulating systems (accounting for observed isotopic exchange (Borle and Uchikawa, 1978) at a low level of in situ Ca where alteration in mitochondrial Ca content results in a new set point. Upon cell stimulation an influx of Ca across the plasma membrane could occur that would elevate Ca_f, in turn elevating total mitochondrial Ca to a level where the mitochondria would contain sufficient Ca to buffer Ca_f at a level higher than the basal level. Consistent with this hypothesis is the report by Borle and Uchikawa (1978) indicating that following addition of parathyroid hormone to cultured kidney cells, Ca entry across the plasma membrane increases 3- to 4-fold. Upon removal of the stimulation, Ca influx falls and Ca efflux remains elevated until Ca_f falls. With the fall in Ca_f, the mitochondria would be expected to release Ca, which would be transported from the cell, and the system would return to the unstimulated state.

On the other hand, if mitochondria buffer Ca_f at a constant level, an influx of Ca across the plasma membrane is likely to be buffered by the mitochondria. If mitochondria function with a saturated efflux pathway, an alteration in Ca_f would have to be accompanied by an alteration in mitochondrial Ca transport pathways. Under basal conditions, the mitochondria could be functioning so that alterations in mitochondrial Ca have little effect on the set point. Stimulation could invoke changes in the mitochondria to alter the set point, and as long as the plasma membrane transporters are operating near V_{max} (Caroni and Carafoli, 1981) they may not be able to effectively lower Ca_f to basal conditions.

Both of these schemes are possible and logic does not eliminate either. In fact, differing cell types may use differing control mechanisms. It is ultimately the plasma membrane transporters that determine whether or not the mitochondria operate to maintain a constant Ca_f (i.e., have a saturated efflux pathway). If the plasma membrane transporters regulate Ca_f at a level below that compatible with a saturated efflux pathway, then the mitochondria lose Ca until they are in equilibrium with Ca_f. There is considerable evidence for differences in plasma membrane Ca transport systems in various cell types (Carafoli and Crompton, 1978). Na-Ca exchange appears to be most prominent in excitable cells. If the presence of Na-Ca exchange resulted in a lower Ca_f, this would cause a lower mitochondrial Ca content and reduce the ability of the mitochondria to buffer Ca_f. However, present data suggest that the K_m for the Ca ATPase is lower than that for the Na-Ca exchanger (Caroni and Carafoli, 1981).

The data at present do not indicate whether or not in situ mitochondria operate with a saturated efflux pathway. There is general agreement that with total mitochondrial Ca contents greater than 4 nmol/mg protein, mitochondria buffer extramitochondrial Ca at a set point of ~0.3 to 0.8 μM (depending on conditions). However, the results of measurements (using various techniques) of in situ mitochondrial Ca contents range so widely (1-40 nmol/mg mitochondrial protein) that they cannot be used to address this controversy. There is also disagreement on the exact relationship between total mitochondrial Ca and the extramitochondrial Ca level. This discrepancy appears to be due to differences in incubation conditions. Increasing Mg and decreasing pH are known to increase the mitochondrial set point (Nicholls, 1978). Joseph and co-workers (1983) also report that although Mg alone increases the set point (0.15 mM Mg gives a set point of 0.7 μM Ca_f compared to 0.2 μM Ca_f without Mg), Mg (0.15 mM) in combination with Mg-ATP (1.6 mM) decreases the set point to 0.5 μM. The authors point out that Becker's data (1980) suggest the effects on the set point of 1 mM Mg and 3 mM ATP are identical to the effects observed without Mg or ATP. The set point measured in the absence of Mg and ATP with mitochondria containing more than 4 nmol Ca/mg protein is ~0.3 μM.

In spite of these differences, there appears to be agreement (see Fig. 4-1) that at a constant level of Ca_f below 0.3 μM the mitochondria will lose Ca via the efflux pathway to a level where the efflux pathway is no longer saturated and that these mitochondria will not buffer Ca_f (see Fig. 4-2). Therefore, in principle, measurements of Ca_f should help determine the suitability of mitochondria for buffering Ca_f. However, this distinction cannot be made given the variability of the data.

In conclusion, at present it is evident that data cannot distinguish between the two possible models put forth for mitochondrial regulations of Ca_f. While there are important differences in these models, both suggest that mitochondria are involved in regulating Ca_f during cell stimulation. The exact resolution to this question will have to await further data and better agreement on levels of Ca_f and in situ mitochondrial Ca content.

ACKNOWLEDGMENTS

I would like to thank Drs. Melvyn Lieberman and Lazaro J. Mandel for their many helpful and stimulating discussions, Connie Ganong for her expert editorial assistance, and Steffani Webb and Gay Blackwell for their able secretarial assistance. This research is supported by a grant from NIH-HL 29687.

REFERENCES

Akerman, K. E. O., and D. G. Nicholls, 1981, Calcium transport by intact synaptosomes, *Eur. J. Biochem.* **115:**67-73.

Akerman, K. E. O., and D. G. Nicholls, 1983, Physiological and bioenergetic aspects of mitochondrial calcium transport, *Rev. Physiol. Biochem. Pharmacol.* **95:**149-201.

Becker, G. L., 1980, Steady state regulation of extramitochondrial Ca^{2+} by rat liver mitochondria. Effects of Mg^{2+} and ATP, *Biochim. Biophys. Acta* **591:**234-239.

Becker, G. L., G. Fiskum, and A. L. Lehninger, 1980, Regulation of free Ca^{2+} by liver mitochondria and endoplasmic reticulum, *J. Biol. Chem.* **255:**9009-9012.

Blaustein, M. P., R. W. Ratzlaff, and E. S. Schweitzer, 1978, Calcium buffering in presynaptic nerve terminal, *J. Gen. Physiol.* **72:**43-66.

Borle, A. B., 1981, Control, modulation, and regulation of cell calcium, *Rev. Physiol. Biochem. Pharmacol.* **90:**13-153.

Borle, A. B., and I. Clark, 1981, Effects of phosphate-induced hyperparathyroidism and parathyroidectomy on rat kidney calcium in vivo, *Am. J. Physiol.* **241:**E136-E141.

Borle, A. B., and T. Uchikawa, 1978, Effects of parathyroid hormone on the distribution and transport of calcium in cultured kidney cells, *Endocrinology* **102:**1725-1732.

Brinley, F. J., Jr., T. Tiffert, and A. Scarpa, 1978, Mitochondria and other calcium buffers of squid axon studied *in situ, J. Gen. Physiol.* **72:**101-127.

Brinley, F. J., Jr., T. Tiffert, A. Scarpa, and L. J. Mullins, 1977, Intracellular calcium buffering capacity in isolated squid axons, *J. Gen. Physiol.* **70:**355-384.

Burgess, G. M., J. S. McKinney, A. Fabiato, B. A. Leslie, and J. W. Putney, Jr., 1983, Calcium pools in saponin-permeabilized guinea pig hepatocytes, *J. Biol. Chem.* **258:**15336-15345.

Bygrave, F. L., 1978, Properties of energy-dependent calcium transport by rat liver microsomal fraction as revealed by initial-rate measurements, *Biochem. J.* **170:**87-91.

Carafoli, E., and M. Crompton, 1978, The regulation of intracellular calcium, *Curr. Top. Memb. Transp.* **10:**151-216.

Caroni, P., and E. Carafoli, 1981, The Ca-pumping ATPase of heart sarcolemma. Characterization, calmodulin dependence and partial purification, *J. Biol. Chem.* **256:**3263-3270.

Charest, R., P. F. Blackmore, B. Berthon, and J. H. Exton, 1983, Changes in free cytosolic Ca^{2+} in hepatocytes following α_1-adrenergic stimulation, *J. Biol. Chem.* **258:**8769-8773.

Coll, K. E., S. K. Joseph, B. E. Corkey, and J. R. Williamson, 1982, Determination of the matrix free Ca^{2+} concentration and kinetics of Ca^{2+} efflux in liver and heart mitochondria, *J. Biol. Chem.* **257:**8696-8704.

Denton, R. M., and J. G. McCormack, 1980, On the role of the calcium transport cycle in heart and other mammalian mitochondria, *FEBS Lett.* **119:**1-8.

Denton, R. M., J. G. McCormack, and N. J. Edgell, 1980, role of calcium ions in the regulation of intramitochondrial metabolism: Effects of Na^+, Mg^{2+}, and ruthenium red on the Ca^{2+} stimulated oxidation of oxoglutarate and on pyruvate dehydrogenase activity in intact rat heart mitochondria, *Biochem. J.* **190:**107-117.

Dipolo, R., J. Requena, F. J. Brinley, Jr., I. J. Mullins, A. Scarpa, and T. Tiffert, 1976, Ionized calcium concentrations in squid axons, *J. Gen. Physiol.* **67:**433-467.

Fiskum, G., and A. L. Lehninger, 1982, Mitochondrial regulation of intracellular calcium, *Calcium Cell Funct.* **2:**39-80.

Hansford, R. G., and F. Castro, 1982, Intramitochondrial and extramitochondrial free calcium ion concentrations of suspensions of heart mitochondria with very low, plausibly physiological, contents of total calcium, *J. Bioenerg. Biomemb.* **14:**361-376.

Hesketh, T. R., G. A. Smith, J. P. Moore, M. V. Taylor, and J. C. Metcalfe, 1983, Free cytoplasmic calcium concentration and the mitogenic stimulation of lymphocytes, *J. Biol. Chem.* **258:**4876-4882.

Joseph, S. K., K. E. Coll, R. H. Cooper, J. S. Marks, and J. R. Williamson, 1983, Mechanisms underlying calcium homeostasis in isolated hepatocytes, *J. Biol. Chem.* **258:**731-741.

Kracke, J., B. F. Sloane, H. Shuman, R. Karp, and A. P. Somlyo, 1980, Electron probe analysis of cultured vascular smooth muscle, *J. Cell. Physiol.* **103:**313-322.

Lee, C. O., A. Taylor, and E. E. Windhager, 1980, Cytosolic calcium ion activity in epithelial cells of *Necturus* kidney, *Nature* **287:**859-861.

Lehninger, A. L., E. Carafoli, and C. S. Rossi, 1967, Energy-linked ion movements in mitochondrial systems, *Adv. Enzymol.* **29:**259-320.

Luft, J. H., 1971, Ruthenium red and violet. II. Fine structured localization in animal tissues, *Anat. Rec.* **121:**369-416.

Mandel, L. J., and E. Murphy, 1984, Regulation of cytosolic free calcium in rabbit proximal renal tubules, *J. Biol. Chem.* **259:**11188-11196.

Marban, E., T. J. Rink, R. W. Tsien, and R. Y. Tsien, 1980, Free calcium in heart muscle at rest and during contraction measured with Ca^{2+}-sensitive microelectrodes, *Nature* **286:**845-850.

McCormack, J. G., and R. M. Denton, 1980, Role of calcium ions in the regulation of intramitochondrial metabolism: Properties of the Ca^{2+} sensitive dehydrogenase within intact uncoupled mitochondria from white and brown adipose tissue of the rat, *Biochem. J.* **190:**95-105.

McCormack, J. G., and P. J. England, 1983, Ruthenium red inhibits the activation of pyruvate dehydrogenase caused by positive inotropic agents in the perfused rat heart, *Biochem. J.* **214:**581-585.

Murphy, E., R. Jacob, and M. Lieberman, 1984, Cytosolic free calcium in chick heart cells: Its role in cell injury, *J. Mol. Cardiol.* **17:**221-231.

Murphy, E., and L. Mandel, 1982, Cytosolic free calcium levels in rabbit proximal kidney tubules, *Am. J. Physiol.* **242:**C124-C128.

Murphy, E., K. Coll, T. L. Rich, and J. R. Williamson, 1980, Hormonal effects on calcium homeostasis in isolated hepatocytes, *J. Biol. Chem.* **255:**6600-6608.

Murphy, E., J. F. Aiton, C. R. Horres, and M. Lieberman, 1983, Calcium elevation in cultured heart cells: Its role in cell injury, *Am. J. Physiol.* **245:**C316-C321.

Nicholls, D. G., 1978, The regulation of extramitochondrial free calcium ion concentration by rat liver mitochondria, *Biochem. J.* **176:**463-474.

Nicholls, D. G., and M. Crompton, 1980, Mitochondrial calcium transport, *FEBS Lett.* **111:**261-268.

O'Doherty, J., and R. J. Stark, 1982, Stimulation of pancreatic acinar secretion: Increases in cytosolic calcium and sodium, *Am. J. Physiol.* **242:**G513-G521.

Pozzan, T., D. P. Lew, C. B. Wollheim, and R. Y. Tsien, 1983, Is cytosolic ionized calcium regulating neutrophil activation? *Science* **221:**1413-1415.

Rasmussen, H., and D. M. Waisman, 1983, Modulation of cell function in the calcium messenger system, *Rev. Physiol. Biochem. Pharmacol.* **95:**111-148.

Requena, J., R. Dipolo, F. J. Brinley, Jr., and L. J. Mullins, 1977, The control of ionized calcium in squid axons, *J. Gen. Physiol.* **70:**329-353.

Scott, I. D., K. E. O. Akerman, and D. G. Nicholls, 1980, Calcium ion transport by intact synaptosomes. Intrasynaptosomal compartmentation and the role of the mitochondrial membrane potential, *Biochem. J.* **192:**873-880.

Shen, A. C., and R. B. Jennings, 1972, Myocardial calcium and magnesium in acute myocardial ischemic injury, *Am. J. Pathol.* **67:**417-440.

Sheu, S., and H. F. Fozzard, 1982, Transmembrane Na^{+} and Ca^{2+} electrochemical gradients in cardiac muscle and their relationship to force development, *J. Gen. Physiol.* **80:**325-351.

Simons, T. J. B., 1981, Measurement of free Ca^{2+} in red blood cells, *J. Physiol.* **318:**38P-39P.

Snowdowne, K. W., and A. B. Borle, 1982, Measurement of intracellular free calcium in monkey kidney cells with aequorin, *Science* **217:**252-254.

Somlyo, A. P., M. Bond, and A. V. Somlyo, 1985, Calcium content of mitochondria and endoplasmic reticulum in liver frozen rapidly in vivo, *Nature* **314:**622-624.

Somlyo, A. V., H. Shuman, and A. P. Somlyo, 1981, Electron probe analysis of cardiac, skeletal, and vascular smooth muscle in microprobe analysis of biological systems, T. H. Hutchinson and A. P. Somlyo, eds., Academic Press, New York, pp. 103-126.

Somlyo, A. P., A. V. Somlyo, and H. Shuman, 1979, Electron probe analysis of vascular smooth muscle. Composition of mitochondria, nuclei, and cytoplasm, *J. Cell. Biol.* **81**:316-333.

Somlyo, A. P., A. V. Somlyo, H. Shuman, B. Sloane, and A. Scarpa, 1978, Electron probe analysis of calcium compartments in cryosections of smooth and striated muscles, *N. Y. Acad. Sci. Ann.* **307**:523-544.

Somlyo, A. V., H. Gonzalez-Serratos, H. Shuman, G. McClellan, and A. P. Somlyo, 1981, Calcium release and ionic changes in the sarcoplasmic reticulum of tetanized muscle: An electron-probe study, *J. Cell Biol.* **90**:577-594.

Steb, H., and I. Schultz, 1983, Regulation of cytosolic free Ca^{2+} concentration in acinar cells of rat pancreas, *Am. J. Physiol.* **245**:G347-G357.

Tsien, R. Y., T. Possan, and T. J. Rink, 1982, Calcium homeostasis in intact lymphocytes: Cytoplasmic free calcium monitored with a new, intracellularly trapped fluorescent indicator, *J. Cell Biol.* **94**:325-334.

Vasington, F. D., and J. V. Murphy, 1961, Active binding of calcium by mitochondria, *Fed. Proc.* **20**:146.

Williamson, J. R., R. H. Cooper, and J. B. Hoek, 1981, Role of calcium in the hormonal regulation of liver metabolism, *Biochim. Biophys. Acta* **639**:243-295.

Williamson, J. R., and E. Murphy, 1980, Calcium homeostasis and compartmentation in liver, in *Alcohol and Aldehyde Metabolizing Systems,* R. G. Thurman, ed., Plenum Press, New York, pp. 671-688.

Williamson, J. R., R. Williams, K. E. Coll, and A. P. Thomas, 1983, Cytosolic free Ca^{2+} concentration and intracellular calcium distribution of Ca^{2+}-tolerant isolated heart cells, *J. Biol. Chem.* **258**:13411-13414.

Wollheim, C. B., and T. Pozzan, 1984, Correlation between cytosolic free Ca^{2+} and insulin release in an insulin-secreting cell line, *J. Biol. Chem.* **259**:2262-2267.

5

Mechanisms of Mitochondrial Damage in Brain Ischemia

Lars Hillered

University of California, San Francisco

INTRODUCTION

Ischemic hypoxia is a condition in which the tissue content of oxygen is decreased as a result of impairment of the blood supply. Tissue hypoxia (or anoxia if complete) can be caused alternatively by reduced oxygen tension (pO_2) in the blood (=hypoxic hypoxia) or by a reduced number of erythrocytes (=anemic hypoxia). In the present context the term *ischemia* will be used as meaning ischemic hypoxia (anoxia).

Several clinical disease states may be complicated by transient episodes of cerebral ischemia. Cardiac arrest and severe shock are examples of conditions that may lead to global ischemic cell damage in the brain, while a thrombus or an embolus may occlude a single vessel and cause a focal ischemic insult (stroke).

Ischemic mitochondrial injury has been implicated as an important factor in the pathogenesis of irreversible ischemic cell damage (Mergner, Smith, and Trump, 1977; Rehncrona, Mela, and Siesjö, 1979; Farber, Chien, and Mittnacht, 1982; Mela, 1979*a,* 1982; Fiskum, 1983). The main topics of the following discussion are the changes in mitochondrial function in response to transient global brain ischemia and the possible mechanisms of such alterations.

PATHOPHYSIOLOGY

As a background some general pathophysiological facts may be recalled. Only events that are relevant to mitochondrial metabolism will be dealt with here. The following literature review is therefore not meant to be a comprehensive but an introductory survey.

Ischemic Thresholds

The reduction in cerebral blood flow (CBF) during ischemia may be complete (as in cardiac arrest) or incomplete (as in shock). This difference has important implications for the metabolic conditions during ischemia.

By means of autoregulation (compensatory modulation of the vascular diameter in response to changes in perfusion pressure) CBF is maintained at a normal level down to mean arterial blood pressures (MABP) of 60 to 70 mmHg (Betz, 1972). Significant signs of tissue hypoxia do not appear, however, until CBF is reduced to 50% of the normal or below, since a compensatory increase in oxygen extraction from the blood will occur (Siesjö, 1978).

When CBF is reduced to certain critical values electrical activity disappears and the membrane potential is disrupted. Electrical activity (electroencephalogram [EEG], evoked potentials) ceases when CBF falls below 30% (Branston et al., 1974), and a release of intracellular K^+ (depolarization) occurs below 10 to 15% of control flow (Astrup et al., 1977). The threshold for cell death seems to be at about the same level (Astrup, Siesjö, and Symon, 1981; Siesjö, 1984*a*).

Energy Metabolism

Normal brain function depends not only upon a constant supply of oxygen (owing to a high metabolic rate and small tissue stores) but also upon a constant supply of glucose from the blood, due partly to the smallness of the tissue stores of glucose and glycogen and partly to the inability of the brain to use other fuel sources (Siesjö, 1978; Kety, 1957). Fatty acids, for example, that are very important for energy metabolism in the heart, seem not to be used by the brain (Owen et al., 1967), perhaps due to the inability of brain mitochondria to take up fatty acids at any significant rate. The contents of carnitine, acetylcarnitine, and carnitine acetyltransferase (the components of the transport system involved in the transport of free fatty acids [FFA] into mitochondria) (Tzagoloff, 1982), have been reported to be very low in brain tissue, and the activity of carnitine acetyltransferase in brain mitochondria has been found to be about 3% of that in heart mitochondria (Marquis and Fritz, 1965). The only alternative substrates that can contribute significantly to energy production in the brain under conditions of reduced glucose availability (e.g., starvation) (Owen et al., 1967) are the ketone bodies β-hydroxybutyrate and acetoacetate. These metabolic features explain the sensitivity of the brain to disturbances in blood flow.

Complete cerebral ischemia has been found to cause energy depletion within a few minutes in several animal species (Siesjö, 1978). In the rat, glucose and phosphocreatine (PCr) were depleted in 1 minute, and virtually no glycogen and only about 10% of the ATP remained after 3 minutes of ischemia (Ljunggren, Shutz, and Siesjö, 1974). In the same study minimal values for energy charge, defined as (ATP + 0.5 ADP)/(ATP + ADP + AMP) (Atkinson, 1968), were obtained after 7.5 minutes. From these data Ljunggren, Shutz, and Siesjö concluded that no useful energy remained after 5 minutes of ischemia.

Clearly, pronounced ischemia rapidly leads to an inhibition of mitochondrial function. Owing to depletion of oxygen, which occurs within seconds after the onset of ischemia (Silver, 1977), mitochondrial respiration will cease, leading to an increased $NADH/NAD^+$ ratio (Ginsberg et al., 1976). At the same time glycolysis is stimulated. As a consequence, lactate accumulation and metabolic acidosis will ensue (Siesjö, 1982, 1984*b*).

During complete brain ischemia, the amount of lactate accumulated is limited by the pre-ischemic tissue contents of glucose and glycogen (Ljunggren, Norberg, and Siesjö, 1974). In normoglycemic rats 10 to 15 μmol g^{-1} wet brain tissue accumulates during 5 to 30 minutes of complete ischemia, and the intracellular pH falls to about 6.5 (Ljunggren, Norberg, and Siesjö, 1974; Rehncrona, Rosén, and Siesjö, 1981).

In contrast, incomplete ischemia will lead to more extensive lactic acidosis as a result of the continued supply of substrate. Lactate levels of 25 to 50 μmol g^{-1} have been reported and the intracellular pH may decrease considerably below 6.0 (Siesjö, 1978).

Ionic Homeostasis

Pronounced ischemia leads to disturbances in ion homeostasis, reflecting energy depletion and membrane depolarization. A few seconds following interruption of CBF a slow increase in extracellular K^+ concentration (K_e^+) occurs (Hansen, 1981). After about 2 minutes there is a steep rise in K_e^+ to about 60 mM (Hansen, 1981; Siemkowicz and Hansen, 1981), accompanied by a decrease in extracellular Na^+ and Cl^- concentration (Na_e^+ and Cl_e^-) and a decrease in extracellular volume (Nicholson, 1979; Hansen, 1981). When K_e^+ reaches a value of 10 to 15 mM there is a sudden marked decrease in extracellular Ca^{2+} concentration (Harris et al., 1981; Harris and Symon, 1984), probably reflecting an opening of voltage dependent Ca^{2+} channels (Berridge, 1979) or, as recently suggested, a transient reversal of the Na^+/Ca^{2+} carrier (Coutinho, Carvalho, and Carvalho, 1984).

These disturbances, which are interpreted as movements of ions and water between extracellular and intracellular compartments, are probably due to failure in the function of the transport systems normally maintaining the ion gradients across the plasmalemma (i.e., the Na^+/K^+ ATPase, the Ca^{2+} ATPase, and the Na^+/Ca^{2+} carrier), since they are directly or indirectly energy-dependent. There are indications that these ionic shifts in the brain are qualitatively different in different cell populations. Thus, influx of Na^+ and Cl^- with a simultaneous swelling occur primarily in glial cells, while Ca^{2+} influx occurs mainly into neurons (Siesjö and Wieloch, 1985).

It is now well established that brain mitochondria possess a sophisticated and independent transport mechanism for both uptake and release of Ca^{2+}. Ca^{2+} uptake occurs through a uniporter driven by the electrochemical gradient over the inner mitochondrial membrane, created by ejection of H^+ by the electron transport chain (Fiskum and Lehninger, 1980; Carafoli, 1982). Efflux of Ca^{2+} occurs mainly via an electroneutral Na^+/Ca^{2+} antiporter (Crompton et al., 1978; Carafoli, 1983). Evidence is emerging that mitochondria play an important role in the regulation of the cytosolic Ca^{2+} concentration in the brain (Mela, 1977; Borle, 1981; Akerman and Nicholls, 1983; Fiskum, 1984).

During ischemia, intracellular conditions will favor a loss of Ca^{2+} from mitochondria and possibly also from other sequestration sites (e.g., the endoplasmic reticulum). Such conditions include a collapsed membrane potential leading to a reversal of the uniporter (Nicholls and Brand, 1980) and increased intracellular Na^+ concentration stimulating the Na^+/Ca^{2+} antiporter. In addition, FFA liberated during ischemia may release Ca^{2+} by acting as uncouplers (Harris, 1977).

Lipid Metabolism

Several investigators have reported on a marked increase in the tissue content of FFA, predominantly arachidonic acid, stearic acid, and palmitic acid during ischemia (Bazán, 1970; Yoshida et al., 1980; Rehncrona et al., 1982). This effect is considered to be caused by a Ca^{2+}-triggered increase in phospholipase activity in conjunction with a decreased phospholipid synthesis owing to ATP depletion (Wieloch and Siesjö, 1982; Siesjö and Wieloch, 1983). The levels of FFA are essentially normalized during the first 30 minutes of recirculation (Yoshida et al., 1980; Rehncrona et al., 1982).

Since brain phospholipase A_2 seems to be located predominantly in the mitochondria (Bazán, 1971; Woelk and Porcellati, 1973; Woelk et al., 1974), the release of FFA during ischemia may have important pathophysiological consequences for mitochondrial metabolism following ischemia.

EXPERIMENTAL MODELS

Before entering a discussion concerning the effects of ischemia and recirculation on mitochondrial function, it seems appropriate to comment on methods for studying brain ischemia in experimental animals and for investigating the function of mitochondria.

Ischemia Models

The brain receives its blood supply mainly through the internal carotid arteries and the vertebral arteries. These two systems are interconnected at the base of the brain in the circle of Willis, allowing an extensive amount of collateral flow both between the two hemispheres and between the two pairs of arteries.

There are other sources of collateral flow to the brain, the most important being the internal mammary arteries through the anterior spinal artery and another being the external carotid arteries via the ophthalmic arteries.

Several different procedures for inducing profound cerebral ischemia have been described in the literature: decapitation, increasing the cerebrospinal fluid (CSF) pressure, inflation of a neck cuff, clamping of supplying arteries, artery clamping in combination with induced systemic hypotension, and cardiac arrest. A few comments about these procedures will follow.

Decapitation is an easy way of inducing complete ischemia, but the recirculation period cannot be studied, and the tissue temperature may be difficult to control during the ischemia (Ozawa et al., 1967; Lazarewicz, Strosznajder, and Gromek, 1972).

Increasing the CSF pressure is a convenient way of achieving reversible complete ischemia. Infusion of an artificial CSF to a pressure above the MABP causes cessation of the CBF. Upon raising the CSF pressure, there is a dramatic increase in MABP, which must be prevented pharmacologically (Neely and Youmans, 1963; Ljunggren, Schutz, and Siesjö, 1974).

Inflation of a neck cuff (the tourniquet method) has been used in the rat (Siemkowicz and Hansen, 1978), the rabbit (Aldrete et al., 1979), and the cat (Nemoto, Hossmann, and Cooper, 1981). Since the vertebral arteries ascend in a bony canal on either side of the upper six cervical vertebrae, it may be difficult to achieve total ischemia. This problem can be overcome by ligating the vertebral arteries in a first session as described by Pulsinelli and Brierley (1979) or by inducing hypotension during the experiment (Siemkowicz and Hansen, 1978; Nemoto, Hossmann, and Cooper, 1981).

A number of procedures for occluding supplying arteries to the brain have been described in the literature. Occlusion of both carotid arteries alone in the rat reduces CBF to about 50% of the control value (Eklof and Siesjö, 1972). To obtain pronounced ischemia, the vertebral arteries also must be occluded (Pulsinelli and Brierley, 1979). However, there is still some collateral flow (already discussed), which may make this procedure less reproducible (Blomquist et al., 1984). This problem has been dealt with by use of an external surgical suture to encircle the neck muscles, thereby reducing collateral flow (Pulsinelli and Duffy, 1983). In larger animals (e.g., dogs) an open-chest model has been used to clamp the ascending aorta and the caval veins for short-term ischemia experiments (Lind, et al., 1975; Steen et al., 1984). Prolonged ischemia has been induced in monkeys with a similar procedure (Hossmann and Zimmermann, 1974).

The Mongolian gerbil has been widely used in brain ischemia research owing to its lack of a complete circle of Willis (Osburne and Halsey, 1975). In this species a profound global brain ischemia consistently follows bilateral occlusion of the common carotid arteries (Ginsberg et al., 1977; Suzuki et al., 1983*a*).

A common way of approaching the problem of collateral flow is to reduce MABP in addition to artery occlusion. This can be achieved either by controlled bleeding (Welsh et al., 1982) or by pharmacologically induced vasodilation (Paschen, Hossmann, and Van den Kerckhoff, 1983), or both. This procedure has been used for many years in rats (Nordström and Siesjö, 1978) and has recently been modified into a long-term survival model (Smith et al., 1984). By means of a combination of pharmacologically induced vasodilation and bleeding in addition to carotid artery occlusion, a well defined, near-complete forebrain ischemia with a rapid onset is consistently produced. Since the vertebral circulation is maintained during ischemia, the brain stem is virtually undamaged, leaving vital functions (e.g., regulation of respiration and blood pressure) intact during the recovery period. This model is currently being used in our laboratory (Hillered, Siesjö, and Arfors, 1984).

Complete cerebral ischemia can also be caused by induced cardiac arrest (or ventricular fibrillation). This procedure has been used mainly in larger animals, for example, dogs (Safar, Stezoski, and Nemoto, 1976) and monkeys (Myers and Yamaguchi, 1977). It is a clinically relevant model, but substantial resources are required to handle the intensive care during and after resuscitation.

Isolation of Brain Mitochondria

Brain mitochondria are very sensitive to ischemia. To obtain good preparations it is therefore crucial that the tissue be rapidly removed following decapitation. In the literature reviewed next essentially two isolation procedures have been employed. They were originally described by Ozawa and co-workers (1966) and by Clark and Nicklas (1970), respectively. The two methods differ mainly with regard to the centrifugation procedures. The Clark-Nicklas technique includes a Ficoll gradient step. Other investigators working on brain mitochondria have used one of these methods with some modifications. The isolation media employed in these studies are based on sucrose or mannitol or a mixture of the two. Ethylenediamine tetraacetate (EDTA) or ethyleneglycol tetraacetate (EGTA) is routinely used to bind calcium that otherwise may influence mitochondrial function.

Bovine serum albumin (BSA) is frequently used to bind FFA that interfere with several energy-linked mitochondrial functions (Lazarewicz, Strosznajder, and Gromek, 1972; Wojtczak, 1976).

In our experiments we have applied a modification of the Clark-Nicklas technique as described by Ginsberg and co-workers (1977; see also Hillered and Ernster, 1983). Rat forebrains were homogenized manually after treatment with a bacterial proteinase (Nagase). The mitochondria were isolated in a mannitol-sucrose-EGTA medium in the presence of BSA.

In rats, with use of the Clark-Nicklas technique, the purity of the mitochondrial preparations (expressed as nmoles of cytochrome a + a_3 per mg of mitochondrial protein) ranges between 0.20 and 0.25, and the protein yield ranges between 7.6 and 9.5 mg of mitochondrial protein per rat brain (Table 5-1). When homogenates from several rats are pooled, the protein yield per brain appears to be lower (Clark and Nicklas, 1970).

According to Clark and Nicklas (1970), their method yields exclusively free mitochondria presumably of glial origin. Electron microscopy reveals that in our hands the preparations contain a mixture of free and synaptosomal (i.e., nerve cell) mitochondria (unpublished results). According to Schutz and co-workers (1973) preparations obtained with the Ozawa technique contain a mixture of free and synaptosomal mitochondria.

Measurement of Mitochondrial Function

The respiration of isolated mitochondria can be measured polarographically (as oxygen consumption) in the presence of suitable substrates. The catalysts of the respiratory chain may occur in varying redox states depending on the experimental conditions (Chance and Williams, 1955). Of the various states of respiration defined by Chance and Williams, two are routinely used in biological experiments as measures of the functional state of isolated mitochondria.

State 3 respiration is the rate of oxygen consumption in the presence of substrate, phosphate, and ADP, while in state 4 the ADP has been converted to ATP. State 4 respiration is often (as in our experiments) measured as the rate of respiration before the addition of ADP.

Respiratory control ratio (RCR) is defined as the ratio of the state 3 to the state 4 rates of oxygen consumption. State 3u is the rate of respiration in the presence of an uncoupler. An uncoupling agent (the classical example being 2,4-dinitrophenol) has the ability to work as a protonophore and thereby abolish the H^+ gradient created by the electron transport chain.

ADP/O ratio is the ratio of the amount of ADP added to the total amount of oxygen consumed (Estabrook, 1967). The maximal ADP/O ratio is 3 with NAD-linked and 2 with FAD-linked substrates.

RCR is considered to be an indicator of the integrity of the isolated mitochondria. Values of RCR with NAD-linked substrates have been reported

Table 5-1. Characteristics of different methods described in the literature for the isolation of brain mitochondria

Species	*Method Cl-Ni/ Ozawa*	*Protein-ase*	*BSA*	*RCR*	*Cytochrome* $a + a_3$	*Protein yield (mg)*	*References*
Rat	Cl-Ni	–	–	7-10	0.25	3.5	Clark and Nicklas, 1970
Rat	Cl-Ni	+	+	6.5	0.23	7.9	Rehncrona, Mela, and Siesjö, 1979
Rat	Cl-Ni	+	+	6.6	0.20	7.6	Hillered and Ernster, 1983
Rat	Cl-Ni	+	+	8-10			Hillered, Ernster, and Siesjö, 1984
Rat	Cl-Ni	+	+	8.5	0.21	8.9	Hillered, Siesjö, and Arfors, 1984
Rat	Ozawa	–	–	11.8			Ozawa et al., 1966
Gerbil	Cl-Ni	+	+	14.7			Ginsberg et al., 1977
Rabbit	Ozawa	+	+	8.9		15.6	Schutz et al., 1973
Guinea pig	Ozawa	–	–	5.4			Lazarewicz, Strosznajder, and Gromek, 1972

Notes: Cl-Ni/Ozawa refers to the originally described methods of Clark and Nicklas (1970) and Ozawa et al. (1966). The bacterial proteinase used is Nagase. BSA = Bovine serum albumin. RCR = respiratory control ratio. Cytochrome $a + a_3$ is in nmol mg^{-1} mitochondrial protein.

to range between 5 and 15 in different studies (Table 5-1). Mitochondrial function measured in vitro as already described allows a first approximation of ischemic alterations. Since the amount of tissue required for the isolation procedure is relatively large, the method is not suitable (at least not in small animals) for the evaluation of regional disturbances in mitochondrial function. Furthermore, such data do not necessarily reflect the actual in vivo activity of the mitochondria. Information concerning mitochondrial function in situ may be deduced from measurements of tissue concentrations of labile phosphates and carbohydrate substrates (Folbergrová, MacMillan, and Siesjö, 1972; Lowry and Passonneau, 1972) or NADH fluorescence in frozen tissue slices (Welsh and Rieder, 1978). The redox state of the cerebral cortex can also be monitored continuously in vivo with fluorescence techniques developed by Chance and co-workers (Chance et al., 1962; Rehncrona, Mela, and Chance, 1979).

MITOCHONDRIAL RESPONSE TO CEREBRAL ISCHEMIA

The following section will deal primarily with the effect of ischemia and recirculation on the function of isolated mitochondria and the subsequent discussion will focus on conceivable mechanisms of such alterations.

Effect of Ischemia

The effect of ischemia on mitochondrial function has been studied extensively in various experimental models using different species and various organs. As discussed by Mela (1982), brain mitochondria appear to be particularly sensitive to ischemia. Table 5-2 summarizes experiments showing the effect of ischemia on state 3 respiration of brain mitochondria, one of the most sensitive parameters of their function, and the approximate time course of these alterations.

In the cited studies, ischemia was accompanied by a 30 to 80% inhibition of state 3 respiration. The onset of this mitochondrial response was usually rapid, occurring within 2 to 5 minutes after decapitation (Ozawa et al., 1967; Lazarewicz, Strosznajder, and Gromek, 1972; Hillered, Siesjö, and Arfors, 1984) and following 15 minutes of two-vessel occlusion combined with bleeding hypotension (Hillered, Siesjö, and Arfors, 1984). The inhibition of state 3 respiration remains in the same range if ischemia is prolonged to 30 to 90 minutes.

There are two exceptions, however. No significant inhibition of state 3 respiration occurred until after 30 minutes of compression ischemia in the rabbit (Schutz et al., 1973) and until after 60 minutes of carotid artery occlusion in the gerbil (Ginsberg et al., 1977). The lack of change in state 3 after

Table 5-2. Summary of different studies on the effect of transient ischemia on brain mitochondrial function

Ischemia Model	*Species*	*Duration of Ischemia (min)*	*% Decrease of* *State 3 Respiration*	*% Decrease of* *ADP/O*	*References*
Decapitation	Rat	2	55	0	Ozawa et al., 1967
		5	75	0	Ozawa et al., 1967
		10	80	20	Ozawa et al., 1967
	Guinea pig	5	45	18	Lazarewicz, Strosznajder, and Gromek, 1972
	Rat	5	55	—	Hillered, Siesjö, and Arfors, 1984
		15	60	—	Hillered, Siesjö, and Arfors, 1984
		30	50	7	Rehncrona, Mela, and Siesjö, 1979
Increased CSF pressure	Rabbit	30	50	—	Schutz et al., 1973
	Rat	30	35-50	11	Rehncrona, Mela, and Siesjö, 1979; Mela, 1979*b*
Bilateral carotid artery occlusion (BCAO)	Gerbil	60-90	30-50	—	Ginsberg et al., 1977
BCAO + hypotension	Rat	15	55	7	Hillered, Siesjö, and Arfors, 1984
		30	65	17	Hillered, Siesjö, and Arfors, 1984
		30	30-35	7	Rehncrona, Mela, and Siesjö, 1979; Mela, 1979*b*
Hemorrhagic shock	Cat	90-180	40-70	—	Mela et al., 1975*a*, 1975*b*

Note: The percentage figures indicate changes in mean values with NAD-linked substrates during ischemia of the given duration compared to controls.

short ischemia periods in these studies may be due to species and/or model differences as suggested by Mela (1979*a*).

As also shown in Table 5-2, hemorrhagic shock (with less pronounced brain ischemia) led to inhibition of state 3 with a later onset.

As just indicated, changes in state 3 respiration is a sensitive indicator of ischemic mitochondrial damage. Addition of an uncoupler such as carbonyl cyanide m-chlorophenylhydrazone (CCCP) did not increase the reduced state 3 respiratory rates (Hillered, Siesjö, and Arfors, 1984), suggesting that an inhibition of the ATP synthase or of the adenine nucleotide translocase that may result from ischemia (Mela, 1979*a*, 1982) was not rate-limiting for state 3 respiration in the isolated mitochondria. It is likely, therefore, that ischemia results in an inhibition of respiration as such.

State 4 respiration has been reported to be unaffected (Ginsberg et al., 1977), slightly decreased (Rehncrona, Mela, and Siesjö, 1979; Lazarewicz, Strosznajder, and Gromek, 1972; Hillered, Siesjö, and Arfors, 1984), or minimally increased (Schutz et al., 1973) by ischemia. As the changes in state 4 respiration, if any, are minimal, the effect of ischemia on RCR is virtually parallel to the decrease in state 3 respiration, a result that by definition is not an uncoupling effect.

ADP/O ratios are largely unaffected by ischemia. As illustrated in Table 5-2, depression of ADP/O occurs later and is much less pronounced compared with the decrease in state 3 respiration. These findings indicate that the remaining capacity of the respiratory chain can be used for ATP synthesis.

It should be stressed here that the ischemic changes just discussed reflect the response not only of mitochondria in different parts of the brain but also in different types of cells. This point is important for two reasons. First, some cell types (i.e., pyramidal cells in layers 3, 5, and 6 of the cerebral cortex and in CA1 and CA4 in the hippocampus, Purkinje cells in the cerebellum, and small to medium sized neurons in the striatum) are selectively vulnerable to ischemia (Siesjö, 1981; Pulsinelli, Brierley, and Plum, 1982). Second, mitochondrial response to incomplete ischemia (measured indirectly by NADH fluorescence in frozen brain sections) is markedly heterogeneous (Welsh, O'Connor, and Langfitt, 1977; Welsh et al., 1978). Thus, evaluation of mitochondrial function after isolation gives only a rough estimate of the mitochondrial response to ischemia. With this important reservation in mind we now proceed to the events of recovery during recirculation.

Effect of Recirculation

There have been only a few direct studies on the recovery of mitochondrial function following transient ischemia. They will therefore be dealt with in greater detail.

One investigation (Rehncrona, Mela, and Siesjö, 1979; Mela, 1979*b*) focused on the recovery of mitochondrial function after 30 minutes of complete (compression) ischemia and after 30 minutes of near-complete forebrain ischemia (two-vessel occlusion + hypotension). The mitochondrial function (state 3 respiration and ADP/O ratio) recovered during 30 minutes of recirculation following complete ischemia, but remained at about the same level as during ischemia (Table 5-2) after near-complete ischemia, suggesting that the difference in recovery was due to the more pronounced lactic acidosis during the latter condition, leading to irreversible mitochondrial damage. One question raised by these results was whether or not the near-complete ischemia was associated with a delayed restitution that was not manifested during the first 30 minutes of recirculation.

That question was explored by Hillered, Siesjö, and Arfors in 1984. Using the same model (two-vessel occlusion + hypotension in the rat), it was found that 30 minutes of ischemia led to a marked decrease in state 3 (and uncoupler-stimulated) respiration that recovered only partly during 60 (-90, not shown) minutes of recirculation (Fig. 5-1*B*); that is, no delayed recovery could be detected. ADP/O ratios were restored when NAD-linked substrates were used, but remained moderately depressed during recirculation with FAD-linked substrates. The reason for this discrepancy is not known.

In the same study the rate of recovery of mitochondrial function was evaluated after a milder ischemic insult (15 minutes in fasted animals) in the same model. State 3 respiration recovered to control level within 30 minutes of recirculation. As early as 5 minutes post-ischemia recovery was extensive, state 3 being only slightly below the control level (Fig. 5-1*A*).

Considering these results together, the following comments seem justified. Although mitochondrial function measured following isolation from whole brain homogenates is a sensitive indicator of cerebral ischemia (Table 5-2), 15 to 30 minutes of ischemia can be tolerated without any signs of irreversible mitochondrial damage. However, two important reservations must be made.

First, the concept of selective vulnerability, as mentioned earlier, implies that preparations from whole brain contain a mixture of damaged and normal mitochondria, and that the proportion of the latter will predominate following insults of short duration. One possible way of overcoming this problem, at least partly, is to isolate mitochondria from tissue samples taken from vulnerable regions of the brain, using large animals. This approach has been used (in cats and dogs) by Mela's group in shock experiments (Mela, 1979*b*; Mela et al., 1975*a*, 1975*b*).

Second, as shown by several investigators (Ito et al., 1975; Kirino, 1982; Pulsinelli, Brierley, and Plum, 1982), maturation of cell damage occurs following ischemia; that is, the final outcome in terms of the number of dead cells is not apparent until after 24 to 72 hours (or more) of recirculation.

Therefore, an initial restitution may not necessarily reflect a permanent return of mitochondrial function. The only mitochondrial study conducted on longer recirculation periods appears to be that reported by Kuwashima and his group (1976). These authors produced a unilateral insult in rats by occlusion of the left carotid artery in combination with systemic hypoxia. A 60 minute insult caused pronounced impairment of state 3 respiration and the ADP/O ratio in mitochondria isolated from the ipsilateral hemisphere, while mitochondria from the contralateral side were normal. The damage was slowly progressive over a period of 72 hours. Although this insult cannot be compared directly with those in the other studies cited, the results suggest that maturation of ischemic mitochondrial damage may also occur.

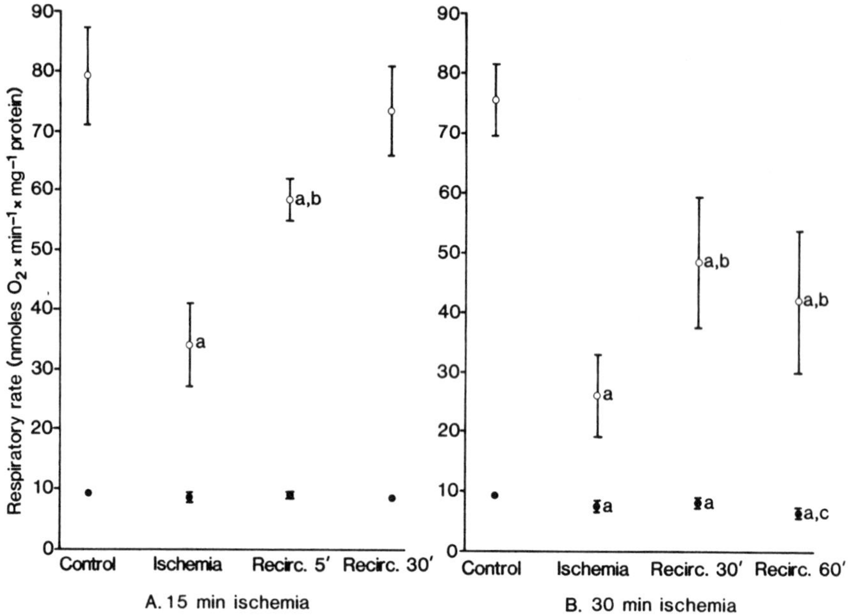

Figure 5-1. Illustration of respiratory rates in brain mitochondria isolated from rats subjected either to 15 minutes of ischemia and 5 or 30 minutes of recirculation (*A*), or to 30 minutes of ischemia and 30 or 60 minutes of recirculation (*B*). State 4 respiration (closed circles) was induced with 9.3 mM malate plus 9.3 mM glutamate as substrate. ADP (0.3–0.5 mM) was used to initiate state 3 respiration (open circles). Mitochondrial protein (0.7–1.1 mg) was added to 0.5 ml of a reaction medium consisting of 150 mM KCl, 10 mM K-phosphate buffer, 1 mM EGTA, pH 7.4. The control groups comprised seven animals and the treated groups six. Values are means with 95% confidence limits indicated. Statistically significant difference from control (a), ischemia (b), and from 30 minutes recirculation group (c) ($p < 0.05$) (after Hillered, Siesjö, and Arfors, 1984).

Since various animal species show neurological (Safar, Stezoski, and Nemoto, 1976; Steen et al., 1983, 1984), morphological (Pulsinelli, Brierley, and Plum, 1982; Suzuki et al., 1983*a;* Smith, Auer, and Siesjö, 1984), and electrophysiological (Suzuki et al., 1983*b*) evidence of irreversible cell damage following 4 to 15 minutes of ischemia, it must be concluded that the respiratory activity of mitochondria isolated from whole-tissue homogenates is not a sensitive indicator of ischemic cell death in the brain. It appears crucial, therefore, that future studies on the role of mitochondrial dysfunction in ischemic cell injury employ methods with higher resolution. Furthermore, isolated mitochondria may be valuable as a model system for studying mechanisms of ischemic cell damage. This aspect of mitochondrial research will be the topic of the following discussion, after a few points concerning the post-ischemic CBF.

MECHANISMS OF MITOCHONDRIAL DAMAGE

CBF During Recirculation

Restoration of an adequate flow to all regions of the brain is a prerequisite for recovery of mitochondrial function and cell viability following transient brain ischemia. In experimental models of complete ischemia, reperfusion may be complicated by the *no-reflow phenomenon* (Ames et al., 1968). This phenomenon is characterized by an inhomogeneous reperfusion, where some regions initially have little or no flow in spite of an adequate perfusion pressure (Kågström, Smith, and Siesjö, 1983*a*).

Incomplete ischemia is not followed by a no-reflow state but by an immediate hyperemia, with a gradual transition to a state of *delayed hypoperfusion* (Kågström, Smith, and Siesjö, 1983*b*). The delayed hypoperfusion could well induce a secondary hypoxic insult that could influence the recovery of mitochondrial function. Both supportive (Paschen, Hossmann, and van den Kerckhoff, 1983) and contradictory (Rehncrona, Mela, and Chance, 1979; Welsh et al., 1982) results are to be found in the literature as to whether or not post-ischemic oxygen delivery is a limiting factor for recovery.

Lactic Acidosis

Lactic acidosis is widely appreciated as an important factor in the pathogenesis of ischemic cell damage in the brain (for reviews see Myers, 1979; Siesjö, 1981; Plum, 1983; Siesjö and Wieloch, 1985). Lactic acid levels of 15 to 25 $\mu mol\ g^{-1}$ (wet tissue weight) during ischemia are of interest in this respect (Siesjö and Wieloch, 1985; Plum, 1983).

It was suggested by Mela (1979*a*) that excessive intracellular acidosis may be an important contributor to irreversible mitochondrial damage caused by ischemia. This hypothesis was based on work done by Nordström, Rehncrona, and Siesjö (1976) and by Rehncrona, Mela, and Siesjö (1979). In these studies it was found that the phosphorylation state of the adenine nucleotide pool and the respiratory activity of isolated mitochondria recovered after complete but not after incomplete ischemia of 30 minutes duration. The only striking difference between these two experimental models was the difference in lactic acidosis during ischemia. In work by Rehncrona, Rosén, and Siesjö in 1980 and 1981, further supportive evidence for Mela's hypothesis was obtained. Following 30 minutes of ischemia induced in rats, the recovery of the cortical energy state varied inversely with the degree of lactic acidosis during the ischemia, provided the tissue lactate concentration exceeded 20 $\mu mol\ g^{-1}$.

To test the hypothesis more directly, an in vitro study was undertaken in our laboratory with the aim of evaluating the effect of lactic acidosis on respiratory activity of isolated brain mitochondria (Hillered, Ernster, and Siesjö, 1984). In agreement with earlier work on mitochondria from other organs (for references see Hillered, Ernster, and Siesjö, 1984), it was found that respiration rate was a function of pH. Compared to the respiratory rate occurring at pH 7.1 (control), a 50% inhibition of state 3 respiration (with NAD-linked substrates) occurred at pH 6.4, a 70% inhibition at pH 6.1, and at pH 5.3 the ADP response was abolished. State 4 respiration was virtually unaffected by acidosis. These experiments were performed in a KCl medium buffered either with phosphate or with bicarbonate and CO_2, the latter to mimic the physiological buffering system. Lowering of the pH by increasing the CO_2-tension in the medium caused similar inhibition of state 3 respiration (Fig. 5-2). Lactate per se did not inhibit respiration.

The results indicate that acidosis, of the same degree as occurring intracellularly during ischemia, severely inhibits mitochondrial function. The implication is that ATP synthesis may be blocked during early recirculation, if the acidosis persists, in spite of the presence of oxygen. Furthermore, it was suggested in the same study that low pH may cause irreversible damage to mitochondria, since state 3 respiration did not recover after prolonged exposure to pH 6.1 in vitro.

In a more recent study (Hillered, Smith, Siesjö, 1985) the effect of lactic acidosis on post-ischemic recovery of mitochondrial function was evaluated in vivo. Ischemia of 15 minutes duration was induced in the same previously mentioned rat model at both high and low tissue lactate concentrations (20 and 10 $\mu mol\ g^{-1}$, respectively) by maintaining different concentrations of plasma glucose prior to ischemia. The results demonstrate that although the functional capacity of the mitochondria recovered somewhat more slowly following high-lactate ischemia (HLI) compared to low-lactate ischemia (LLI), recovery of the phosphorylation state of the adenine nucleotide pool was at least as rapid and extensive in HLI as in LLI animals. These data indicate that the effect of the more pronounced intracellular acidosis during early recirculation following HLI (pH about 6.5 at 5 minutes post-ischemia as against 6.8 in LLI animals) did not significantly impair the in vivo activity of the mitochondria. A possible explanation for these findings is that mitochondria normally respire at a submaximal level in vivo. Considering these results together with those of the studies cited earlier, it is suggested that the negative effect of lactic acidosis on mitochondrial function is related

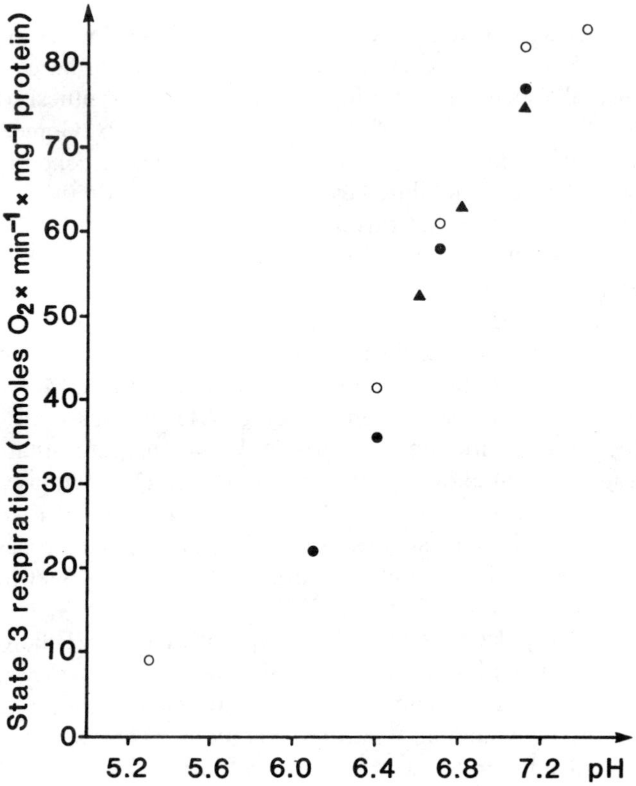

Figure 5-2. Comparison of the effects of in vitro lactic acidosis in a reaction medium (150 mM KCl, 1 mM EGTA, pH 7.1 or 7.4) buffered with bicarbonate and CO_2 (closed circles) or with a phosphate buffer (as in Fig. 5-1; open circles) and of hypercapnia (triangles) on the oxygen consumption rate in state 3 respiration (nmols O_2 min^{-1} mg^{-1} protein) of isolated rat brain mitochondria. Each point represents the mean value of 4 to 6 preparations (after Hillered, Ernster, and Siesjö, 1984).

both to the degree of lactic acidosis and to the duration of ischemia and that an effect of acidosis on mitochondrial function may be detrimental only if (and when) oxygen supply is restricted (e.g., during incomplete ischemia).

This conclusion would account for the finding that 30 minutes of incomplete ischemia, induced in fed animals (allowing excessive lactic acidosis to develop), is more detrimental to mitochondrial function than 30 minutes of complete ischemia (Rehncrona, Mela, and Siesjö, 1979), while there is no difference in the mitochondrial activities between HLI and LLI animals following 15 minutes of complete ischemia (Hillered, Smith, and Siesjö, 1985).

Ionic Shifts

Mitochondrial Ca^{2+} uptake competes with ATP synthesis since both processes utilize the same energy source (Lehninger, 1982). Furthermore, mitochondrial Ca^{2+} uptake seems to occur in preference to ATP synthesis (Rossi and Lehninger, 1964; Vercesi, Reynafarje, and Lehninger, 1978; Gómez-Puyou et al., 1980). It has recently been shown that in brain mitochondria, ADP-stimulated respiration is inhibited by the uptake of Ca^{2+} (in the 50–100 nmol mg^{-1} protein range) and that this inhibition is not reversed unless Ca^{2+} is released from the mitochondria (Roman, Clark, and Swanson, 1981; Nowicki, MacKenzie, and Young, 1982). These findings suggest that Ca^{2+} accumulated intracellularly during ischemia may block ATP synthesis at an early stage post-ischemia and thereby prolong the period of energy depletion.

In an attempt to test this hypothesis, we have found that Mn^{2+}, taken up together with Ca^{2+}, prevents the inhibition of ATP synthesis in isolated rat brain mitochondria (Hillered et al., 1983). It was suggested that Ca^{2+} and Mn^{2+}, within the mitochondria, interact with the ATP synthase inhibitor protein in such a way that Mn^{2+} promotes and Ca^{2+} inhibits the release of the inhibitor bound to the ATP synthase during the Ca^{2+} uptake process. Subsequent experiments with submitochondrial particles from beef heart have supported this conclusion (Ernster, 1984).

We have recently also tested the above hypothesis in vivo (Hillered, Siesjö, and Arfors, 1984; Hillered, Smith, and Siesjö, 1985). In the first study, relatively little Ca^{2+} had accumulated in the mitochondria after 15 minutes of ischemia (about 5 nmol mg^{-1} protein) and recovery of respiratory activity of the isolated mitochondria was extensive during the first 5 minutes of recirculation. Since intramitochondrial Ca^{2+} may be lost during isolation of the mitochondria, a second study was undertaken, in which the phosphorylation state of the adenine nucleotide pool was also measured, as a reflection of the mitochondrial function in vivo. In that study, the tissue content of Ca^{2+} was determined (in addition to the mitochondrial Ca^{2+} content). The Ca^{2+} that accumulated intracellularly during ischemia did not interfere with a rapid recovery of either the energy state or the mitochondrial respiratory activity.

It was concluded that calcium homeostasis is rapidly restored after 15 minutes of ischemia, without disturbing the early recovery of mitochondrial function.

Some results reported by Ginsberg and co-workers (1977) and Schutz with co-workers (1973) suggested that the impairment of mitochondrial function during ischemia may partly be due to a loss of mitochondrial Mg^{2+}. In a subsequent study, Rehncrona, Mela, and Siesjö (1979) found that the pattern of mitochondrial alterations during ischemia and recirculation was not influenced by additions of Mg^{2+} to the mitochondria in vitro. Our own data showed that a loss of mitochondrial Mg^{2+} did occur during ischemia (about 25%), but also that mitochondrial function recovered after ischemia without normalization of the Mg^{2+} content (Hillered, Siesjö, and Arfors, 1984).

Membrane Alterations

Rehncrona, Mela, and Siesjö (1979) found a decrease in the total yield of mitochondrial protein during 30 minutes of ischemia. Mitochondria with gross structural changes may have been lost during the centrifugation procedure, owing to a change in density. However, there was no statistically significant decrease in the yield of cytochrome $a + a_3$. The change in protein yield could therefore also be explained by varying degrees of nonmitochondrial contamination.

Schutz and co-workers (1973) found no reduction of the protein yield during 5 to 40 minutes of ischemia. We also observed no decrease in the yield of either cytochrome $a + a_3$ or protein during 15 or 30 minutes of ischemia, suggesting that no such selection phenomenon occurred (Hillered, Siesjö, and Arfors, 1984).

Ischemia has been reported to induce alterations in mitochondrial membrane phospholipids. Decapitation ischemia of 5 minutes duration has been found to be associated with a 20% decrease in the mitochondrial phospholipid content (Majewska, Strosznajder, and Lazarewicz, 1978). The largest individual changes occurred in the phosphatidyl choline and phosphatidyl ethanolamine fractions. Pilot experiments in our laboratory have revealed a decrease in the cardiolipin concentration (by about 20%) following 15 minutes of ischemia (unpublished). Similar changes have been found in mitochondria from other organs exposed to ischemia (Smith et al., 1980). As discussed by Fiskum (1983), such alterations of membrane phospholipids may have a profound influence on mitochondrial function and possibly explain at least some of the ischemia-induced mitochondrial damage. This suggestion would also account for the finding already discussed that mitochondrial function recovered following 15 minutes of ischemia if the mitochondria were isolated after 5 or 30 minutes of recirculation, while the inhibition of mitochondrial function persisted during isolation (in spite of the presence of oxygen, normalized pH, etc.) if no recirculation was allowed. This finding agrees with the idea of a repair process during early recirculation in vivo that does not take place in vitro.

Oxygen Radicals and Lipid Peroxidation

The concept of oxygen radical-induced reactions as important factors in the pathogenesis of ischemic cell damage in the brain has received increasing attention over the last few years (Flamm et al., 1978; Siesjö, 1981; Mela, 1982; Siesjö and Wieloch, 1985; White, Wiegenstein, and Winegar, 1984; Chan and Fishman, 1985). To test this concept at the mitochondrial level, a study was conducted to evaluate the influence of oxygen radicals on mitochondrial function (Hillered and Ernster, 1983). Following in vitro exposure to hypoxanthine and xanthine oxidase (a commonly used radical-generating system) in the presence of chelated iron, state 3 respiration was markedly inhibited. It was concluded that this effect (probably located at the NADH-dehydrogenase level of the respiratory chain) was caused by the hydroxyl radical formed in a Fenton-type reaction catalyzed by iron, mobilized from intracellular stores by the superoxide radical (Fig. 5-3). There are other possible radical-generating systems in vivo following ischemia (e.g., arachidonic acid metabolism) that, in principle, can replace hypoxanthine and xanthine oxidase in the scheme

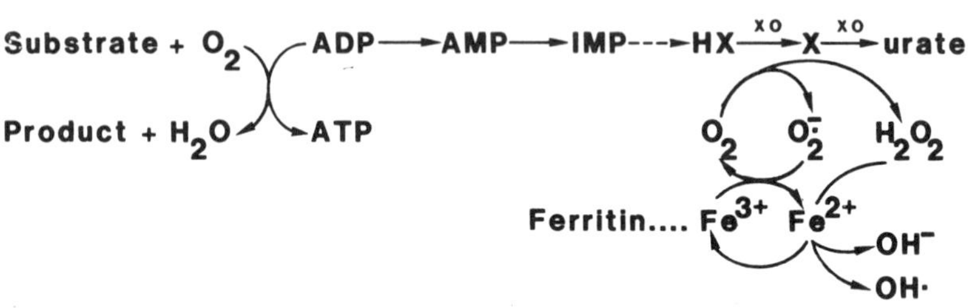

Figure 5-3. Schematic illustration of the conceivable sequence of reactions leading to post-ischemic radical formation. When respiration is inhibited by lack of oxygen during ischemia, ATP synthesis from ADP is diminished and AMP formation is enhanced, leading to an increased formation of hypoxanthine, which, owing to lack of oxygen, accumulates. Upon recirculation, hypoxanthine is oxidized to xanthine and urate (by xanthine oxidase), with the concomitant formation of $O_2^{\bar{\cdot}}$ and H_2O_2. In conjunction with iron, mobilized from intracellular stores (e.g., ferritin-Fe^{3+}) through reduction by $O_2^{\bar{\cdot}}$, OH· will be formed. This burst of oxygen radicals may overcome the capacity of the physiological defense systems, such as superoxide dismutase, catalase, and glutathione peroxidase, and thereby cause cell damage (after Hillered and Ernster, 1983).

of Figure 5-3. A similar mechanism may contribute to the irreversible mitochondrial damage observed during recirculation following prolonged ischemia (Rehncrona, Mela, and Siesjö, 1979; Hillered, Ernster, and Arfors, 1984).

The increase in FFA during ischemia may have adverse effects on mitochondrial function for several reasons. FFA may impair the respiratory activity of brain mitochondria (Lazarewicz, Strosznajder, and Gromek, 1972); this direct effect seems to be complex, however, and may vary qualitatively depending on the concentration of FFA to which the mitochondria are exposed (Wojtczak, 1976).

Chan and Fishman (1980) have reported that FFA added to brain cortical slices induced edema and production of superoxide radicals. Chan, Yurko, and Fishman (1982) found that xanthine oxidase, xanthine, and ferric iron caused edema and lipid peroxidation in various subcellular fractions, including mitochondria, in the same in vitro system. Mitochondrial phosphatidyl choline has been found to be a good substrate for lipid peroxidation in vitro (Sharma, 1977) and lipid peroxidation has been detected in the mitochondrial phospholipid fraction during brain ischemia (Majewska, Strosznajder, and Lazarewicz, 1978).

During recirculation the accumulated pool of FFA may be oxidized by cyclo-oxygenase and lipoxygenase. Formation of cyclo-oxygenase as well as lipoxygenase products has been reported to occur in the brain during recirculation (Gaudet and Levine, 1980; Moskowitz et al., 1984). Apart from the properties of the prostaglandins (and the leukotrienes) as inflammatory mediators (that may contribute to post-ischemic cell damage per se), some reactions involved in the oxidation of FFA during recirculation may be important generators of oxygen radicals and initiators of lipid peroxidation in vivo (Chan and Fishman, 1980; Wieloch and Siesjö, 1982; Wolfe, 1982).

In conclusion, although a considerable amount of indirect evidence has been reported, the importance of oxygen radicals and lipid peroxidation in the pathogenesis of ischemic cell damage in the brain remains to be elucidated.

SUMMARY AND CONCLUSIONS

This chapter has dealt with the mitochondrial response to transient brain ischemia and recirculation and has discussed the mechanisms of such alterations. As an introduction, some relevant pathophysiological and methodological aspects have been considered. The following main conclusions have been drawn:

Mitochondrial function measured as the respiratory activity of isolated mitochondria is rapidly inhibited following the onset of pronounced brain ischemia.

During recirculation mitochondrial function recovers after 15 minutes of near-complete ischemia or 30 minutes of complete ischemia. Thirty minutes of near-complete ischemia causes irreversible mitochondrial damage.

Acidosis induced in vitro by lactic acid markedly inhibits mitochondrial respiratory activity. Pronounced lactic acidosis induced during 15 minutes of ischemia does not cause irreversible mitochondrial damage and does not slow down the rate of recovery of the phosphorylation state of the adenine nucleotide pool, indicating that the capacity of the mitochondrial respiration is sufficient to allow rephosphorylation in spite of a partial inhibition caused by the low intracellular pH during the initial stage of recirculation. It is concluded that the negative effect of lactic acidosis on mitochondrial function as reported in other studies is related to both the degree of acidosis and the duration of ischemia.

Although in vitro data indicate that mitochondrial Ca^{2+} sequestration interferes with ATP synthesis, Ca^{2+} accumulated intracellularly during ischemia, does not inhibit ATP synthesis early after ischemia, indicating that Ca^{2+} homeostasis is rapidly restored following ischemia of 15 minutes duration.

Indirect evidence is presented that membrane alterations may constitute the changes in mitochondrial function caused by ischemia and that oxygen radical mechanisms may be involved in the pathogenesis of irreversible mitochondrial damage.

Mitochondrial function measured as the respiratory activity of mitochondria isolated from whole-brain homogenates during the first stage of recirculation is not a sensitive indicator of irreversible ischemic cell damage. However, isolated mitochondria is a valuable model system for in vitro studies on mechanisms of such injury.

Methods with higher resolution are required to further elucidate the role of mitochondrial dysfunction in ischemic cell injury.

ACKNOWLEDGMENTS

The author is very grateful to Dr. L. Ernster and Dr. B. K. Siesjö for constructive criticism. The study was supported by the Swedish Society of Medical Sciences, the Trygg-Hansa Foundation, and the Volvo Company.

REFERENCES

Akerman, K. E. O., and D. G. Nicholls, 1983, Physiological and bioenergetic aspects of mitochondrial calcium transport, *Rev. Physiol. Biochem. Pharmacol.* **95:**149-201.

Aldrete, J. A., F. Romo-Salas, L. Jankovsky, and Y. Franatovic, 1979, Effect of pretreatment with thiopental and phenytoin on postischemic brain damage in rabbits, *Crit. Care Med.* **7:**466-470.

Ames, A. III, R. L. Wright, M. Kowada, J. M. Thurston, and G. Majno, 1968, Cerebral ischemia II. The no-reflow phenomenon, *Am. J. Pathol.* **52:**437-453.
Astrup, J., B. K. Siesjö, and L. Symon, 1981, Thresholds in cerebral ischemia. The ischemic penumbra, *Stroke* **12:**723-725.
Astrup, J., L. Symon, N. M. Branston, and N. A. Lassen, 1977, Cortical evoked potential and extracellular K^+ and H^+ at critical levels of brain ischemia, *Stroke* **8:**51-57.
Atkinson, D. E., 1968, The energy charge of the adenylate pool as a regulatory parameter. Interaction with feedback modifiers, *Biochemistry* **7:**4030-4034.
Bazán, N. G., 1970, Effects of ischemia and electroconvulsive shock on free fatty acid pool in the brain, *Biochim. Biophys. Acta.* **218:**1-10.
Bazán, N. G., Jr., 1971, Phospholipases A_1 and A_2 in brain subcellular fractions, *Acta Physiol. Latinoam.* **21:**101-106.
Berridge, M. J., 1979, Modulation of nervous activity by cyclic nucleotides and calcium, in *The Neurosciences: Fourth Study program,* F. O. Schmitt and F. G. Worden, eds., MIT Press, Cambridge, Mass., pp. 873-889.
Betz, E., 1972, Cerebral blood flow: Its measurement and regulation, *Physiol. Rev.* **52:**595-630.
Blomqvist, P., H. Mabe, M. Ingvar, and B. K. Siesjö, 1984, Models for studying long-term recovery following forebrain ischemia in the rat. 1: Circulatory and functional effects of four-vessel occlusion, *Acta Neurol. Scand.* **69:**376-384.
Borle, A. B., 1981, Control, modulation, and regulation of cell calcium, *Rev. Physiol. Biochem. Pharmacol.* **90:**13-153.
Branston, N. M., L. Symon, H. A. Crockard, and E. Pasztor, 1974, Relationship between the cortical evoked potential and local cortical blood flow following acute middle cerebral artery occlusion in the baboon, *Exp. Neurol.* **45:**195-208.
Carafoli, E., 1982, The transport of calcium across the inner membrane of mitochondria, in *Membrane Transport of Calcium,* E. Carafoli, ed., Academic Press, New York, pp. 109-139.
Carafoli, E., 1983, Uptake and release of calcium by mitochondria: Recent developments and role in intracellular calcium regulation, in *Pathobiology of Cell Membranes,* vol. III, B. F. Trump and A. U. Arstila, eds., Academic Press, New York, pp. 185-200.
Chan, P. H., and R. A Fishman, 1980, Transient formation of superoxide radicals in polyunsaturated fatty acid-induced brain swelling, *J. Neurochem.* **35:**1004-1007.
Chan, P. H., and R. A. Fishman, 1985, Free fatty acids, oxygen free radicals, and membrane alterations in brain ischemia and injury, in *Cerebrovascular Diseases,* F. Plum and W. Pulsinelli, eds., Raven Press, New York, pp. 161-171.
Chan, P. H., M. Yurko, and R. A. Fishman, 1982, Phospholipid degradation and cellular edema induced by free radicals in brain cortical slices, *J. Neurochem.* **38:**525-531.
Chance, B., and G. R. Williams, 1955, A method for the localization of sites for oxidative phosphorylation, *Nature* **176:**250-254.
Chance, B., P. Cohen, F. Jobsis, and B. Schoener, 1962, Intracellular oxidation-reduction states *in vivo.* The microfluorometry of pyridine nucleotide gives a continuous measurement of the oxidation state, *Science* **137:**499-507.
Clark, J. B., and W. J. Nicklas, 1970, The metabolism of rat brain mitochondria, *J. Biol. Chem.* **245:**4724-4731.
Coutinho, O. P., C. A. M. Carvalho, and A. P. Carvalho, 1984, Calcium uptake related to K^+-depolarization and Na^+/Ca^+ exchange in sheep brain synaptosomes, *Brain Res.* **290:**261-271.
Crompton, M., R. Moser, H. Lüdi, and E. Carafoli, 1978, The interrelations between the transport of sodium and calcium in mitochondria of various mammalian tissues, *Eur. J. Biochem.* **82:**25-31.
Eklöf, B., and B. K. Siesjö, 1972, The effect of carotid artery ligation upon the blood flow and energy state of the rat brain, *Acta Physiol. Scand.* **86:**155-165.
Ernster, L., 1984, Mechanism and regulation of mitochondrial ATP synthesis, *Curr. Top. Cell Reg.* **24:**313-334.

Estabrook, R. W., 1967, Mitochondrial respiratory control and the polarographic measurement of ADP:O ratios, in *Methods in Enzymology,* vol. 10, R. W. Estabrook and M. E. Pullman, eds., Academic Press, New York, pp. 41-47.

Farber, J. L., K. R. Chien, and S. Mittnacht, Jr., 1982, The pathogenesis of irreversible cell injury in ischemia, *Am. J. Pathol.* **102:**271-281.

Fiskum, G., 1983, Involvement of mitochondria in ischemic cell injury and in regulation of intracellular calcium, *Am. J. Emerg. Med.* **2:**147-153.

Fiskum, G., 1984, Physiological aspects of mitochondrial calcium transport, in *Calcium and Its Role in Biology,* H. Sigel, ed., Metal Ions in Biological Systems, vol. 17, Marcel Dekker, New York, pp. 187-214.

Fiskum, G., and A. L. Lehninger, 1980, The mechanisms and regulation of mitochondrial Ca^{2+} transport, *Fed. Proc.* **39:**2432-2436.

Flamm, E. S., H. B. Demopoulos, M. L. Seligman, R. G. Poser, and J. Ransohoff, 1978, Free radicals in cerebral ischemia, *Stroke* **9:**445-447.

Folbergrová, J., V. MacMillan, and B. K. Siesjö, 1972, The effect of moderate and marked hypercapnia upon the energy state and upon the cytoplasmic $NADH/NAD^+$ ratio of the rat brain, *J. Neurochem.* **19:**2497-2505.

Gaudet, R. J., and L. Levine, 1980, Effect of unilateral common carotid artery occlusion on levels of prostaglandins D_2, $F_{2\alpha}$ and 6-keto-prostaglandin $F_{1\alpha}$ in gerbil brain, *Stroke* **11:**648-652.

Ginsberg, M. D., M. Reivich, S. Frinak, and K. Harbig, 1976, Pyridine nucleotide redox state and blood flow of the cerebral cortex following middle cerebral artery occlusion in the cat, *Stroke* **7:**125-131.

Ginsberg, M. D., L. Mela, K. Wrobel-Kuhl, and M. Reivich, 1977, Mitochondrial metabolism following bilateral cerebral ischemia in the gerbil, *Ann. Neurol.* **1:**519-527.

Gómez-Puyou, M. T., M. Gavilanes, A. Gómez-Puyou, and L. Ernster, 1980, Control of activity states of heart mitochondrial ATPase. Role of the proton-motive force and Ca^{2+}, *Biochim. Biophys. Acta* **592:**396-405.

Hansen, A. J., 1981, Extracellular ion concentrations in cerebral ischemia, in *The Application of Ion Selective Microelectrodes,* T. Zeuthen, ed., Elsevier/North-Holland Biomedical Press, Amsterdam, pp. 239-254.

Harris, E. J., 1977, The uptake and release of calcium by heart mitochondria, *Biochem. J.* **168:**447-456.

Harris, R. J., and L. Symon, 1984, Extracellular pH, potassium, and calcium activities in progressive ischaemia of rat cortex, *J. Cereb. Blood Flow Metabol.* **4:**178-186.

Harris, R. J., L. Symon, N. M. Branston, and M. Bayhan, 1981, Changes in extracellular calcium activity in cerebral ischaemia, *J. Cereb. Blood Flow Metabol.* **1:**203-209.

Hillered, L., and L. Ernster, 1983, Respiratory activity of isolated rat brain mitochondria following *in vitro* exposure to oxygen radicals, *J. Cereb. Blood Flow Metabol.* **3:**207-214.

Hillered, L., L. Ernster, and K.-E. Arfors, 1984, Brain ischemia and oxygen radicals, in *Advances in Brain Ischemia Research,* A. Bes, P. Braquet, R. Paoletti, and B. K. Siesjö, eds., Elsevier/ North-Holland Biomedical Press, Amsterdam, pp. 293-300.

Hillered, L., L. Ernster, and B. K. Siesjö, 1984, Influence of *in vitro* lactic acidosis and hypercapnia on respiratory activity of isolated rat brain mitochondria, *J. Cereb. Blood Flow Metabol.* **4:**430-437.

Hillered, L., B. K. Siesjö, and K.-E. Arfors, 1984, Mitochondrial response to transient forebrain ischemia and recirculation in the rat, *J. Cereb. Blood Flow Metabol.* **4:**438-446.

Hillered, L., M.-L. Smith, and B. K. Siesjö, 1985, Lactic acidosis and recovery of mitochondrial function following forebrain ischemia in the rat. *J. Cereb. Blood Flow Metabol.* **5:**259-266.

Hillered, L., P. M. Muchiri, K. Nordenbrand, and L. Ernster, 1983, Mn^{2+} prevents the Ca^{2+}-induced inhibition of ATP synthesis in brain mitochondria, *FEBS Lett.* **154:**247-250.

Hossmann, K.-A., and V. Zimmermann, 1974, Resuscitation of the monkey brain after 1 H complete ischemia. 1. Physiological and morphological observations, *Brain Res.* **81:**59-74.

Ito, U., M. Spatz, J. T. Walker, Jr., and I. Klatzo, 1975, Experimental cerebral ischemia in Mongolian gerbils. I. Light microscopic observations, *Acta Neuropathol.* **32:**209-223.

Kågström, E., M.-L. Smith, and B. K. Siesjö, 1983*a,* Local cerebral blood flow in the recovery period following complete cerebral ischemia in the rat, *J. Cereb. Blood Flow Metabol.* **3:**170-182.

Kågström, E., M.-L. Smith, and B. K. Siesjö, 1983*b,* Recirculation in the rat brain following incomplete ischemia, *J. Cereb. Blood Flow Metabol.* **3:**183-192.

Kety, S. S., 1957, The general metabolism of the brain *in vivo,* in *The Metabolism of the Nervous System,* D. Richter, ed., Pergamon Press, London, pp. 221-237.

Kirino, T., 1982, Delayed neuronal death in the gerbil hippocampus following ischemia, *Brain Res.* **239:**57-69.

Kuwashima, J., B. Fujitani, K. Nakamura, T. Kadokawa, K. Yoshida, and M. Shimizu, 1976, Biochemical changes in unilateral brain injury in the rat: A possible role of free fatty acid accumulation, *Brain Res.* **110:**547-557.

Lazarewicz, J. W., J. Strosznajder, and A. Gromek, 1972, Effects of ischemia and exogenous fatty acids on the energy metabolism in brain mitochondria, *Acad. Pol. Sci. Bull.* **20:**599-606.

Lehninger, A. L., 1982, *Principles of Biochemistry,* Worth Publishers, Inc, New York.

Lind, B., J. Snyder, S. Kampschulte, and P. Safar, 1975, A review of total brain ischemia models and original experiments on clamping the aorta, *Resuscitation* **4:**19-31.

Ljunggren, B., K. Norberg, and B. K. Siesjö, 1974, Influence of tissue acidosis upon restitution of brain energy metabolism following total ischemia, *Brain Res.* **77:**173-186.

Ljunggren, B., H. Schutz, and B. K. Siesjö, 1974, Changes in energy state and acid-base parameters of the rat brain during complete compression ischemia, *Brain Res.* **73:**277-289.

Lowry, O. H., and J. V. Passonneau, 1972, *A Flexible System of Enzymatic Analysis,* Academic Press, New York.

Majewska, M. D., J. Strosznajder, and J. Lazarewicz, 1978, Effect of ischemic anoxia and barbiturate anesthesia on free radical oxidation of mitochondrial phospholipids, *Brain Res.* **158:**423-434.

Marquis, N. R., and I. B. Fritz, 1965, The distribution of carnitine, acetylcarnitine, and carnitine acetyltransferase in rat tissue, *J. Biol. Chem.* **240:**2193-2196.

Mela, L., 1977, Mechanisms and physiological significance of calcium transport across mammalian mitochondrial membranes, *Curr. Top. Membr. Transp.* **9:**321-366.

Mela, L., 1979*a,* Mitochondrial function in cerebral ischemia and hypoxia: Comparison of inhibitory and adaptive responses, *Neurol. Res.* **1:**51-63.

Mela, L., 1979*b,* Reversibility of mitochondrial metabolic response to circulatory shock and tissue ischemia, *Circ. Shock Suppl.* **1:**61-67.

Mela, L., 1982, Mitochondrial function in shock, ischemia, and hypoxia, in *Pathophysiology of Shock, Anoxia, and Ischemia,* R. A. Cowley and B. F. Trump, eds., Williams & Wilkins, Baltimore, pp. 84-95.

Mela, L., K. Harbig, A. Kovach, and M. Reivich, 1975*a,* Mitochondrial function in shock, *Stroke* **6:**236-237.

Mela, L., W. Crowe, K. Harbig, K. Wrobel-Kuhl, and A. Kovach, 1975*b,* Inhibition of brain mitochondrial function and changes of tissue H^+ and K^+ concentration in hemorrhagic shock, *Surg. Forum* **26:**51-53.

Mergner, W. J., M. W. Smith, and B. F. Trump, 1977, Studies on the pathogenesis of ischemic cell injury XI. P/O ratio and acceptor control, *Virchows Arch. Sec. B: Cell Pathol.* **26:**17-26.

Moskowitz, M. A., K. J. Kiwak, K. Hekimian, and L. Levine, 1984, Synthesis of compounds with properties of leukotrienes C_4 and D_4 in gerbil brains after ischemia and recirculation, *Science* **224:**886-888.

Myers, R. E., 1979, A unitary theory of causation of anoxic and hypoxic brain pathology, in *Cerebral Hypoxia and Its Consequences,* S. Fahn, J. N. Davis, and L. P. Rowland, eds., Advances in Neurology, vol. 26, Raven Press, New York, pp. 195-213.

Myers, R. E., and S. Yamaguchi, 1977, Nervous system effects of cardiac arrest in monkeys, *Arch. Neurol.* **34:**65-74.

Neely, W. A., and J. R. Youmans, 1963, Anoxia of canine brain without damage, *Am. Med. Assoc. J.* **183:**1085-1087.

Nemoto, E. M., K.-A. Hossmann, and H. K. Cooper, 1981, Post-ischemic hypermetabolism in cat brain, *Stroke* **12:**666-676.

Nicholls, D. G., and M. D. Brand, 1980, The nature of the calcium ion efflux induced in rat liver mitochondria by the oxidation of endogenous nicotinamide nucleotides, *Biochem. J.* **188:**113-118.

Nicholson, C., 1979, Brain-cell microenvironment as a communication channel, in *The neurosciences: Fourth Study Program,* F. O. Schmitt and F. G. Worden, eds., MIT Press, Cambridge, Mass., pp. 457-476.

Nordström, C.-H., S. Rehncrona, and B. K. Siesjö, 1976, Restitution of cerebral energy state after complete and incomplete ischemia of 30 min duration, *Acta Physiol. Scand.* **97:**270-272.

Nordström, C.-H., and B. K. Siesjö, 1978, Effects of phenobarbital in cerebral ischemia. Part I: Cerebral energy metabolism during pronounced incomplete ischemia, *Stroke* **9:**327-335.

Nowicki, J. P., E. T. MacKenzie, and A. R. Young, 1982, Brain ischemia, calcium and calcium antagonists, *Pathol. Biol.* **30:**282-288.

Osburne, R. C., and J. H. Halsey, Jr., 1975, Cerebral blood flow. A predictor of recovery from ischemia in the gerbil, *Arch. Neurol.* **32:**457-461.

Owen, O. E., A. P. Morgan, H. G. Kemp, J. M. Sullivan, M. G. Herrera, and G. F. Cahill, 1967, Brain metabolism during fasting, *J. Clin. Invest.* **46:**1589-1595.

Ozawa, K., K. Seta, H. Takeda, K. Ando, H. Handa, and C. Araki, 1966, On the isolation of mitochondria with high respiratory control from rat brain, *J. Biochem.* **59:**501-510.

Ozawa, K., K. Seta, H. Araki, and H. Handa, 1967, The effect of ischemia on mitochondrial metabolism, *J. Biochem.* **61:**512-514.

Paschen, W., K.-A. Hossmann, and W. van den Kerckhoff, 1983, Regional assessment of energy-producing metabolism following prolonged complete ischemia of cat brain, *J. Cereb. Blood Flow Metabol.* **3:**321-329.

Plum, F., 1983, What causes infarction in ischemic brain? The Robert Wartenberg lecture, *Neurology* **33:**222-233.

Pulsinelli, W. A., J. B. Brierley 1979, A new model of bilateral hemispheric ischemia in the unanesthetized rat, *Stroke* **10:**267-272.

Pulsinelli, W. A., J. B. Brierley, and F. Plum, 1982, Temporal profile of neuronal damage in a model of transient forebrain ischemia, *Ann. Neurol.* **11:**491-498.

Pulsinelli, W. A., and T. E. Duffy, 1983, Regional energy balance in rat brain after transient brain ischemia, *J. Neurochem.* **40:**1500-1503.

Rehncrona, S., L. Mela, and B. Chance, 1979, Cerebral energy state, mitochondrial function, and redox state measurements in transient ischemia, *Fed. Proc.* **38:**2489-2492.

Rehncrona, S., L. Mela, and B. K. Siesjö, 1979, Recovery of brain mitochondrial function in the rat after complete and incomplete cerebral ischemia, *Stroke* **10:**437-446.

Rehncrona, S., I. Rosén, and B. K. Siesjö, 1980, Excessive cellular acidosis: an important mechanism of neuronal damage in the brain? *Acta Physiol. Scand.* **110:**435-437.

Rehncrona, S., I. Rosén, and B. K. Siesjö, 1981, Brain lactic acidosis and ischemic cell damage: 1. Biochemistry and neurophysiology, *J. Cereb. Blood Flow Metabol.* **1:**297-311.

Rehncrona, S., B. Westerberg, B. Åkesson, and B. K. Siesjö, 1982, Brain cortical fatty acids and phospholipids during and following complete and severe incomplete ischemia, *J. Neurochem.* **38:**84-93.

Roman, I., A. Clark, and P. D. Swanson, 1981, The interaction of calcium transport and ADP phosphorylation in brain mitochondria, *Membr. Biochem.* **4:**1-9.

Rossi, C. S., and A. L. Lehninger, 1964, Stoichiometry of respiratory stimulation, accumulation of Ca^{2+} and phosphate, and oxidative phosphorylation in rat liver mitochondria, *J. Biol. Chem.* **239:**3971-3980.

Safar, P., W. Stezoski, E. M. Nemoto, 1976, Amelioration of brain damage after 12 minutes cardiac arrest in dogs, *Arch. Neurol.* **33:**91-95.

Schutz, H., P. R. Silverstein, M. Vapalahti, D. A. Bruce, L. Mela, and T. W. Langfitt, 1973, Brain mitochondrial function after ischemia and hypoxia, *Arch. Neurol.* **29:**408-416.

Sharma, O. P., 1977, Peroxidation of rat brain mitochondrial lipids, *J. Neurochem.* **28:**1377-1379.

Siemkowicz, E., and A. J. Hansen, 1978, Clinical restitution following cerebral ischemia in hypo-, normo-, and hyperglycemic rats, *Acta Neurol. Scand.* **58:**1-8.

Siemkowicz, E., and A. J. Hansen, 1981, Brain extracellular ion composition and EEG activity following 10 minutes ischemia in normo- and hyperglycemic rats, *Stroke* **12:**236-240.

Siesjö, B. K., 1978, *Brain Energy Metabolism,* Wiley, New York.

Siesjö, B. K., 1981, Cell damage in the brain: A speculative synthesis, *J. Cereb. Blood Flow Metabol.* **1:**155-185.

Siesjö, B. K., 1982, Lactic acidosis in the brain: Occurrence, triggering mechanisms and pathophysiological importance, in *Metabolic Acidosis,* Ciba Foundation Symposium 87, Pitman Books Ltd., London, pp. 77-100.

Siesjö, B. K., 1984*a,* Cerebral circulation and metabolism, *J. Neurosurg.* **60:**883-908.

Siesjö, B. K., 1984*b,* Brain acid-base metabolism in health and disease, in *Advances in Brain Ischemia Research,* A. Bes, P. Braquet, R. Paoletti, and B. K. Siesjö, eds., Elsevier/North-Holland Biomedical Press, Amsterdam, pp. 157-165.

Siesjö, B. K., and T. Wieloch, 1983, Fatty acid metabolism and the mechanisms of ischemic brain damage, in *Cerebrovascular diseases,* M. Reivich and H. I. Hurtig, eds., Raven Press, New York, pp. 251-268.

Siesjö, B. K., and T. Weiloch, 1985, Brain injury: Neurochemical aspects, in *Central Nervous System Trauma—Status Report,* J. Povlischock and D. Becker, eds., William Byrd Press, Inc., Richmond, Va., pp. 513-532.

Silver, I. A., 1977, Changes in pO_2 and ion fluxes in cerebral hypoxia-ischemia, in *Tissue Hypoxia and Ischemia,* M. Reivich, ed., Plenum Press, New York, pp. 299-312.

Smith, M.-L., R. N. Auer, and B. K. Siesjö, 1984, The density and distribution of ischemic brain injury in the rat following two to ten minutes of forebrain ischemia, *Acta Neuropathol.* **64:**319-332.

Smith, M. W., Y. Collan, M. W. Kahng, and B. F. Trump, 1980, Changes in mitochondrial lipids of rat kidney during ischemia, *Biochim. Biophys. Acta.* **618:**192-201.

Smith, M.-L., G. Bendek, N. Dahlgren, I. Rosén, T. Wieloch, and B. K. Siesjö, 1984, Models for studying long-term recovery following forebrain ischemia in the rat. 2. A two vessel occlusion model, *Acta Neurol. Scand.* **69:**385-401.

Steen, P. A., L. A. Newberg, J. H. Milde, and J. D. Michenfelder, 1983, Nimodipine improves cerebral blood flow and neurologic recovery after complete cerebral ischemia in the dog, *J. Cereb. Blood Flow Metabol.* **3:**38-43.

Steen, P. A., L. A. Newberg, J. H. Milde, and J. H. Michenfelder, 1984, Cerebral blood flow and neurologic outcome when nimodipine is given after complete cerebral ischemia in the dog, *J. Cereb. Blood Flow Metabol.* **4:**82-87.

Suzuki, R., T. Yamaguchi, T. Kirino, F. Orzi, and I. Klatzo, 1983*a,* The effects of 5-minute ischemia in Mongolian gerbils: I. Blood-brain barrier, cerebral blood flow, and local cerebral glucose utilization changes, *Acta Neuropathol.* **60:**207-216.

Suzuki, R., T. Yamaguchi, C.-L. Li, and I. Klatzo, 1983*b,* The effects of 5-minute ischemia in Mongolian gerbils: II. Changes of spontaneous neuronal activity in cerebral cortex and CA1 sector of hippocampus, *Acta Neuropathol.* **60:**217-222.

Tzagoloff, A., 1982, *Mitochondria (Cellular Organelles),* P. Siekevitz, ed., Plenum Press, New York.

Vercesi, A., B. Reynafarje, A. L. Lehninger, 1978, Stoichiometry of H^+ ejection and Ca^{2+} uptake coupled to electron transport in rat heart mitochondria, *J. Biol. Chem.* **253:**6379-6385.

Welsh, F. A., M. J. O'Connor, and T. W. Langfitt, 1977, Regions of cerebral ischemia located by pyridine nucleotide fluorescence, *Science* **198:**951-953.

Welsh, F. A., and W. Rieder, 1978, Evaluation of *in situ* freezing of cat brain by NADH fluorescence, *J. Neurochem.* **31:**299-309.

Welsh, F. A., M. D. Ginsberg, W. Rieder, and W. W. Budd, 1978, Diffuse cerebral ischemia in the cat: II. Regional metabolites during severe ischemia and recirculation, *Ann. Neurol.* **3:**493-501.

Welsh, F. A., M. J. O'Connor, V. R. Marcy, A. J. Spatacco, and R. L. Johns, 1982, Factors limiting regeneration of ATP following temporary ischemia in cat brain, *Stroke* **13:**234-242.

White, B. C., J. G. Wiegenstein, and C. D. Winegar, 1984, Brain ischemic anoxia: Mechanisms of injury, *Am. Med. Assoc. J.* **251:**1586-1590.

Wieloch, T., and B. K. Siesjö, 1982, Ischemic brain injury: The importance of calcium, lipolytic activities, and free fatty acids, *Pathol. Biol.* **30:**269-277.

Woelk, H., and G. Porcellati, 1973, Subcellular distribution and kinetic properties of rat brain phospholipases A_1 and A_2, *Hoppe-Seyler's Z. Physiol. Chem.* **354:**90-100.

Woelk, H., K. Pieler-Ichikawa, L. Binaglia, G. Goracci, and G. Porcellati, 1974, Distribution and properties of phospholipases A_1 and A_2 in synaptosomes and subsynaptosomal fractions of rat brain, *Hoppe-Seyler's Z. Physiol. Chem.* **355:**1535-1542.

Wojtczak, L., 1976, Effect of long-chain fatty acids and acyl-CoA on mitochondrial permeability, transport, and energy-coupling processes, *J. Bioenerg. Biomembr.* **8:**293-311.

Wolfe, L. S., 1982, Eicosanoids: Prostaglandins, thromboxanes, leukotrienes, and other derivatives of carbon-20 saturated fatty acids, *J. Neurochem.* **38:**1-14.

Yoshida, S., S. Inoh, T. Asano, K. Sano, M. Kubota, H. Shimazaki, and N. Ueta, 1980, Effect of transient ischemia on free fatty acids and phospholipids in the gerbil brain, *J. Neurosurg.* **53:**323-331.

6

Effects of Ischemia and Calcium Antagonists on Ca^{2+} Transport by Heart Mitochondria

Jeanie B. McMillin-Wood and Paul E. Wolkowicz

Baylor College of Medicine

INTRODUCTION

A primary involvement of Ca^{2+}-transporting organelles in the etiology of ischemic injury to the heart has been suggested by numerous investigations on isolated mitochondria and sarcoplasmic reticulum. However, the biochemical process leading to irreversible cellular injury remains elusive. A general consensus exists that disruption of the mechanisms controlling cellular Ca^{2+} homeostatis either is central to onset of irreversibility or reflects one of a constellation of events that are hallmarks of myocyte death.

The clinically relevant form of cardiac ischemia, due to either coronary artery spasm or occlusion, is that in which blood flow, and thus oxygen delivery to the tissue, is greatly diminished. This phenomenon represents a further complication to the simple lack of oxygen. For example, low-flow ischemia compromises both the delivery of energy-producing substrates (e.g., glucose) to the heart and the washout of potentially toxic compounds that accumulate as end products of anaerobic metabolism. The major metabolites known to be deleterious to cardiac function are lactate, which decreases intracellular pH, and natural amphiphiles, for example, fatty acylcarnitine, −CoA and lysophosphatides, which may perturb membrane bilayer integrity and inhibit a variety of membrane-bound ion transport ATPases essential to the maintenance of cellular ion gradients. Roles for each of these metabolites in ischemic dysfunction and injury have been interpreted for the most part with respect to disturbances in intracellular ion homeostasis, with particular emphasis on Ca^{2+} (Katz and Messineo, 1981). The physiological and biochemical parameters associated with the various stages of ischemia are summarized as follows.

CELLULAR RESPONSE TO ISCHEMIA

The initial cellular response to decreased oxygen delivery and diminished cellular washout is a stimulation of anaerobic glycogenolysis and glycolysis with accumulation of lactate in the sarcoplasm. These sequellae of metabolic events are accompanied by an immediate decrease in muscle contractility although greater than 75% of the cellular ATP stores are maintained. The rapid decrease in contractile function most likely occurs as a result of lactic acidosis due to competition of H^+ for Ca^{2+} binding sites on troponin C of the actomyosin complex. Impairment of Ca^{2+} uptake by the sarcoplasmic reticulum may also be a consequence of the increased H^+ concentration (Fabiato and Fabiato, 1978). However, diminished Ca^{2+} transport by isolated reticulum vesicles has also been demonstrated independent of pH effects in early ischemia (Feher, Briggs, and Hess, 1980).

The combination of H^+ competition for Ca^{2+} binding sites and impaired Ca^{2+} binding and uptake by the sarcoplasmic reticulum results in elevations of cytosolic free Ca^{2+} concentrations. These changes require Ca^{2+} accumulation by the mitochondria in order to maintain cytoplasmic free Ca^{2+} homeostasis. Since the cytoplasmic phosphorylation potential is fairly well maintained during the early response to ischemia, it is likely that the mitochondrion would be able to play an effective role in the transport of Ca^{2+} by the potential-dependent uniport mechanism in an effort to buffer small rises in the cystolic Ca^{2+} concentration. Studies from this laboratory (Wood et al., 1979) have described significant decreases in phosphorylating respiration in mitochondria isolated within 5 to 10 minutes following diminished oxygen delivery. While these changes represent only minor alterations (25%) in the V_{max} of ADP-stimulated oxygen consumption, they may signal the beginnings of the next stage in ischemic injury.

During the period up to 30 to 60 minutes following coronary artery ligation, the ATP levels in the cell fall to 30% of normal values (Neely et al., 1973) and the total levels of nucleotides decrease as a consequence of purine nucleotide degrading pathways (Berne and Rubio, 1974). This second stage of ischemic damage is reversible because reperfusion of the ischemic area produces a gradual return to normal physiological function of the heart. Although mitochondria isolated from the ischemic myocardium during this reversible period demonstrate a capability for continued oxidative phosphorylation and Ca^{2+} sequestration, these rates are markedly diminished. Moreover, the ability of isolated mitochondria to retain accumulated Ca^{2+} is severely compromised, with spontaneous Ca^{2+} release occurring presumably by reversal of the uniport pathway (Pinsky et al., 1981). The exact nature of the events leading to increased Ca^{2+} efflux from mitochondria is not known, but several hypotheses have been proposed from experimentation in vivo.

One hypothesis relates to the dramatic decrease in ATP concentration in the cell occurring at this time. Since the fall in the cytoplasmic phosphorylation potential is thought to be in near equilibrium with the mitochondrial membrane potential (Williamson, 1979), ischemic mitochondria would be expected to demonstrate a markedly decreased capacity for Ca^{2+} accumulation in vitro. Decreases in total matrix adenine nucleotides during the period of reversible ischemia have also been correlated with decreases in rates of oxidative phosphorylation and inability to retain accumulated Ca^{2+} in vitro (Sordahl and Stewart, 1980; Asimakis and Sordahl, 1981). It has also been demonstrated that palmitoyl-CoA, a metabolite that accumulates in myocardial ischemia and inhibits the adenine nucleotide translocase from both the matrix and cytosolic aspects of the mitochondrial inner membrane, produces a premature discharge of mitochondrial Ca^{2+} (Asimakis and Sordahl, 1977). Other work has suggested that alterations in mitochondrial membrane permeability may be a reflection of Ca^{2+}-activated phospholipase A_2 activity in the inner membrane (Palmer and Pfeiffer, 1981).

Although the exact sequence of events leading to mitochondrial Ca^{2+} release is not established, the net result of the fall in mitochondrial Ca^{2+} transport activity is further elevation of the intracellular free Ca^{2+}. The duration of this increased intracellular free Ca^{2+} set point is dependent upon operation of Ca^{2+} extrusion mechanisms at the sarcolemma because ultimate regulation of intracellular Ca^{2+} must reside with the ability of cell plasma membrane Ca^{2+} pumps to maintain Ca^{2+} homeostasis. Inhibition of the calmodulin-dependent sarcolemmal Ca^{2+} ATPase in vivo may be caused by a decrease in available ATP as well as by H^+ competition for Ca^{2+} binding sites on calmodulin (Williamson, Woodrow, and Scarpa, 1975; Caroni and Carafoli, 1981). Decreased activity of the Na^+-Ca^{2+} exchange in the cell plasma membrane has also been demonstrated in sarcolemmal vesicles isolated from ischemic heart (Bersohn, Philipson, and Fukishima, 1982). The reversible phase of ischemic injury is associated with only modest changes in myocardial Ca^{2+} (Shen and Jennings, 1972), although gradual elevation of cytoplasmic free Ca^{2+} is physiologically apparent by elevations in end-diastolic pressure during ischemia (Hearse, Garlick, and Humphrey, 1977). The lack of an appreciable change in intracellular free Ca^{2+} during reversible ischemia is most likely due to maintenance of sarcolemmal integrity so that uncontrolled Ca^{2+} influx into the cell does not occur upon reperfusion. Therefore, cell viability is preserved upon restoration of oxygen and substrate, and physiological function eventually returns to a normal state.

Alterations in the structural integrity of the sarcolemma are associated with the final stage of ischemic damage. This stage is defined by irreversible injury and is characterized by a lack of mechanical recovery of the heart upon reperfusion. Dramatic changes in myocyte ultrastructure are observed, including myofibrillar disruption, intracellular lipofuscin granules, clearing of the mitochondrial matrix, swelling of the mitochondria and the presence of dense granules that contain Ca^{2+} in the matrix (Ganote et al., 1975). It is likely that these changes reflect the consequences of a primary alteration in sarcolemmal permeability, although the factors responsible for the latter change are speculative. However, it can be reasonably assumed that disruption of phospholipid bilayer packing may play a role in ischemia-induced sarcolemmal alterations. One hypothesis to account for cellular membrane disruption suggests that an increase in Ca^{2+}-activated phospholipase activity produces lysophosholipids in the absence of energy-dependent re-esterification (Corr, Lee, and Sobel, 1981). Alterations in phospholipid composition, loss of phospholipids, and perturbations of membrane integrity by membrane-active amphiphiles (e.g., palmitoylcarnitine) have also been proposed to play a role in sarcolemmal permeability increases (Katz and Messineo, 1981). It is especially noteworthy that investigators have observed inhibition of lysolecithin acyltransferase activity by palmitoylcarnitine (Gross, Drisdel, and Sobel, 1983), a metabolite that accumulates to millimolar concentrations in ischemia (Idell-Wenger, Grotyohann, and Neely, 1978). The sarcolemmal membrane may be particularly susceptible to phospholipid modifications by virtue of its high phospholipid to protein composition in comparison with intracellular membranes.

Manifestation of irreversible ischemic damage may not be apparent until blood flow is restored. The rapid deterioration of function upon reperfusion has been attributed to uncontrolled Ca^{2+} entry into the cell. The increase in passive Ca^{2+} accumulation is thought to produce Ca^{2+} overload in the mitochondria at the expense of oxidative ATP production (Hearse, 1977). Ca^{2+} cycling across the inner mitochondrial membrane under ischemic conditions results in an irreversible collapse of the mitochondrial membrane potential. It is unlikely that the sarcolemmal Ca^{2+} pumps are able to maintain Ca^{2+} homeostasis during Ca^{2+} overload. Thus, intracellular free Ca^{2+} rises precipitously with subsequent activation of Ca^{2+} dependent proteases and lysosomal enzymes (Beinlich et al., 1981; Beckman et al., 1982).

CALCIUM CHANNEL BLOCKERS

Due to the apparent intimate involvement of cellular Ca^{2+} homeostatic control in the progression of ischemic damage to cell death, many recent studies have been concerned with therapeutic mechanisms to prevent Ca^{2+} overload. Verapamil (α-isopropyl-[(N-methyl-N-homo veratryl)-aminopropyl]-3,4-dimethoxyphenylacetonitrile hydrochloride) is a prototype drug of a class of compounds designated as organic Ca^{2+} antagonists. Other drugs included under this general group are D600 (the methoxy derivative of verapamil), nifedipine, and diltiazem. Electrophysiological studies have demonstrated that these compounds are inhibitors of the cell membrane slow Ca^{2+} current (Fleckenstein et al., 1969) and are effective in the prevention of Ca^{2+} overload that occurs with reperfusion of jeopardized myocardium (Henry et al., 1977). Increased Ca^{2+} entry during ischemia and reperfusion probably does not occur on the slow Ca^{2+} channel, and it is likely that this increase in sarcolemmal permeability to Ca^{2+} is a manifestation rather than a primary cause of injury. Therefore, it is not likely that the mode of action of the Ca^{2+} antagonists to prevent ischemic damage is solely via impedence of slow Ca^{2+} entry.

One possible explanation for myocardial protection by Ca^{2+} antagonists is the negative inotropic action of these compounds that reduces cardiac work and energy demand of the compromised myocardium, thus preserving myocardial ATP stores (Fleckenstein, 1977). However, recent work by Hamm and Opie (1983) suggests that the protective effects of these Ca^{2+} antagonists is not achieved by decreased heart work because diltiazem depresses neither cardiac output nor left ventricular function, but does diminish irreversible damage. Another possible mode of action is a decrease of coronary vascular resistance with a resultant increase in collateral blood flow to the ischemic area. However, protective effects of verapamil, nifedipine, and diltiazem have been demonstrated in the isolated working rat heart preparation that has no peripheral circulation (Hamm and Opie, 1983).

Due to their lipophilic properties, Ca^{2+} antagonists are likely to accumulate in the myocardial cell. Therefore, it is possible that these drugs have additional, subcellular effects on the energy metabolism of the ischemic cell. We have demonstrated that 60 minute normothermic ischemic arrest of canine hearts in the presence of KCl and verapamil produces significantly higher functional recovery of these hearts upon reperfusion (Pinsky et al., 1981). This recovery, in turn, is associated with a preservation of control rates of mitochondrial phosphorylating respiration. However, it is difficult from these types of studies to assign a primary role for mitochondrial protection in ischemic reversibility. Moreover, it has been demonstrated that diltiazem at low concentrations has no direct effect on mitochondrial oxidative phosphorylation (Vághy et al., 1981). In this regard, both verapamil and diltiazem inhibit phosphate-induced swelling of isolated mitochondria, the latter being associated with related inhibition of ADP-stimulated respiration (Vághy et al., 1981). Also, neither verapamil nor D-600 affects Ca^{2+} binding or release, or the associated ATPase activity in cardiac sarcoplasmic reticulum (Watanabe and Besch, 1974), nor do any reports exist for an effect of these compounds on Ca^{2+} transport in isolated sarcolemmal vesicles.

These essentially negative reports concerning a possible subcellular action of Ca^{2+} antagonists contrast with more recent reports from Vághy and co-workers (1982) and our own laboratory (Wolkowicz et al., 1983) of an apparent direct effect of these drugs on Na^{+}-Ca^{2+} exchange in the cardiac mitochondrial inner membrane. In energized mitochondria, Ca^{2+} uptake proceeds by the potential-dependent uniport pathway. Net efflux must occur by extrusion of Ca^{2+} against its electrochemical potential gradient. The influx and efflux pathways are distinguished on the basis of differential sensitivities to ruthenium red, lanthanides, and Na^{+}. Na^{+} causes a complete release of mitochondrial Ca^{2+} in the presence of ruthenium red (Crompton, Capano, and Carafoli, 1976), a release believed to occur by a Na^{+}-Ca^{2+} antiport mechanism, although Na^{+}-H^{+}, Ca^{2+}-H^{+} exchange pathways may act in concert to produce net Ca^{2+} extrusion. The Ca^{2+} antagonist drugs, including diltiazem, prenylamine, fendiline, nifedipine, and verapamil, are the only known specific inhibitors of mitochondrial Na^{+}-induced Ca^{2+} release.

ISCHEMIA AND MITOCHONDRIAL CALCIUM MOVEMENTS

With cessation of aerobic metabolism, the products of the nonoxidative pathway of cardiac fatty acid metabolism accumulate. The sites of the synthesis of palmitoyl-CoA and palmitoylcarnitine in the heart are the outer (palmitoyl-CoA) and inner (both compounds) mitochondrial membranes. Therefore, it is presumed that significant concentrations of these metabolites are present on the membrane and at the membrane interface. Because of the amphipathic nature of acyl esters, it is not surprising that they interact with all cellular membranes to affect associated enzymatic processes. We have demonstrated significant inhibition of sarcolemmal Na^{+}, K^{+}-ATPase activity (Wood et al., 1977) as well as Ca^{2+} uptake into sarcoplasmic reticulum vesicles (Pitts et al., 1978) by palmitoylcarnitine at concentrations well below those measured in the ischemic myocardium. The levels of amphipaths required to produce disruption of these membrane-associated processes are further decreased by vesicle preincubation with the fatty acyl esters. The time-dependence of inhibition is a reflection of the detergent properties of the amphiphiles. At high concentrations, palmitoylcarnitine inhibits both Na^{+}, K^{+}-ATPase and Ca^{2+}, Mg^{2+}-ATPase activities, most likely through protein denaturation effects.

Both palmitoyl-CoA and palmitoylcarnitine have concentration-dependent inhibition effects on cardiac mitochondrial Ca^{2+} transport at amphiphile and protein concentrations that assure absence of micellar effects (Wolkowicz and McMillin-Wood, 1980*a*). Calcium uptake (53 μM) by heart mitochondria (2.8 mg) was followed in the presence of 100 μM tetramethylmurexide (518 – 542 nm). NADH absorption was measured in the absence of dye at 370 –

340 nm. Addition of low concentrations of either palmitoyl-CoA or palmitoylcarnitine to control preparations of Ca^{2+}-loaded cardiac mitochondria produces a net release of the accumulated Ca^{2+} (Fig. 6-1 *A, C*). The time required to observe net Ca^{2+} release depends on the concentration of amphiphile added, and release is sensitized by the presence of phosphate. In the absence of phosphate, little or no Ca^{2+} release is observed even at high acyl ester concentrations. Ca^{2+} release induced by either palmitoyl-CoA or palmitoylcarnitine is accompanied by pyridine nucleotide oxidation (Fig. 6-1 *B, D*), large amplitude swelling, uncoupling of respiration, and a loss in membrane potential. When ruthenium red (an inhibitor of potential-dependent Ca^{2+}uptake) is added to Ca^{2+}-loaded mitochondria, palmitoyl-CoA and palmitoylcarnitine addition produces a slowed Ca^{2+} release, with an absence of swelling and pyridine nucleotide oxidation (Fig. 6-1 *A*, ± RR).

In addition to producing premature Ca^{2+} release from heart mitochondria, the presence of palmitoyl CoA or palmitoylcarnitine is associated with slowed rates of Ca^{2+} accumulation in cardiac mitochondria. The inhibited uptake follows an uncompetitive inhibition pattern with K_I values of 0.4 μM and 20 μM for palmitoyl-CoA and palmitoylcarnitine, respectively (Wolkowicz and McMillin-Wood, 1980*a*). The described anomalies in Ca^{2+} transport kinetics mediated by these amphiphiles appear to have a unique dependence upon phosphate in the medium. No Ca^{2+} efflux is observed if either lactate or acetate is substituted for phosphate (Wolkowicz and McMillin-Wood, 1980*b*). Furthermore, the presence of N-ethylmaleimide, which inhibits the phosphate:OH^- exchange, inhibits palmitoyl-CoA mediated Ca^{2+} release (Wolkowicz and McMillin-Wood, 1980*b*). Acyl-CoA esters increase carrier-independent permeability of the membrane to phosphate and have been shown to release inhibition of mitochondrial swelling produced by mersalyl in isotonic ammonium phosphate (Wojtczak, 1976).

Therefore these types of studies can partially reproduce the sensitivity of isolated ischemic mitochondria to Ca^{2+}-induced injury. It is likely that the metabolic events occurring in the ischemic cell, (i.e., fatty acyl ester and phosphate accumulation) may play a role in changes in cellular Ca^{2+} homeostasis. A third important consequence of ischemia is cellular acidosis, which, it has been proposed, modifies Ca^{2+}-dependent reactivities and Ca^{2+} transport events in the heart. It has been demonstrated by numerous laboratories that isolated liver mitochondria transport and retain Ca^{2+} less effectively as the pH of the suspending medium is decreased (Nicholls, 1978; Wolkowicz and McMillin-Wood, 1981). Acid pulses produce release of accumulated Ca^{2+} from liver mitochondria (Akerman, 1978). These results have been interpreted in terms of a Ca^{2+}/nH^+ exchange reaction (Fiskum and Cockrell, 1978) that represents a passive, electroneutral pathway for Ca^{2+} efflux.

It is not likely that a comparable proton-coupled pathway for Ca^{2+} efflux

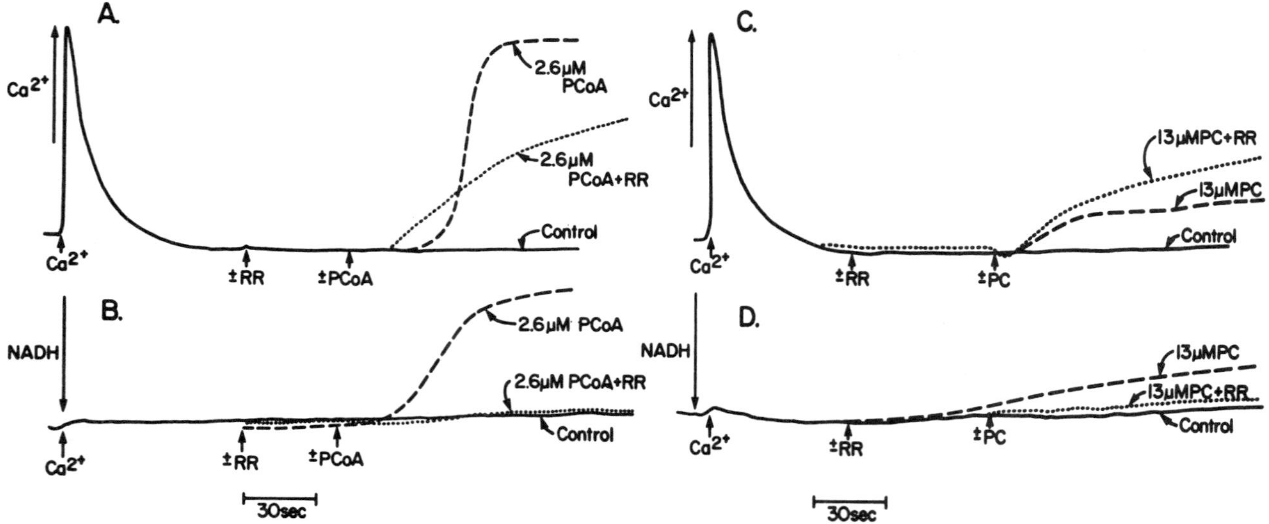

Figure 6-1. Effects of palmitoyl-CoA and palmitoyl carnitine on Ca^{2+} efflux and pyridine nucleotide oxidation in the presence and absence of ruthenium red (from Wolkowicz and McMillin-Wood, 1980*a*).

exists in heart mitochondria (Jurkowitz and Brierley, 1982). The existence of a Na^+-Ca^{2+} exchange pathway in mitochondria from excitable tissues represents the mechanism for Ca^{2+} extrusion from heart mitochondria. A decrease in the pH of the medium slightly elevates basal Ca^{2+} efflux without altering the Na^+ stimulatable component (Harris and Heffron, 1982). Coupling of Na^+ movements to a Na^+:H^+ antiport has been proposed (Crompton, Capano, and Carafoli, 1976). However, rather than imparting a pH sensitivity to Ca^{2+} efflux in heart mitochondria, this exchanger has been postulated by others (Jurkowitz and Brierley, 1982) to protect against deleterious effects of low pH on mitochondrial Ca^{2+} homeostasis in the heart by dissipating ΔpH, which is the driving force for phosphate entry.

Jurkowitz and Brierley (1982) have demonstrated that a drop in pH produces rapid and extensive loss of Ca^{2+} from heart mitochondria. However, although the exact sequence of events leading to Ca^{2+} efflux from heart mitochondria cannot be specified, evidence suggests that a simple electroneutral H^+-Ca^{2+} exchange is not involved. Phosphate is required for release to occur, as well as an oxidized NAD(P)H pool in the matrix (Jurkowitz and Brierley, 1982). Heart mitochondria are further sensitized to a H^+ induced release of Ca^{2+} by depletion of endogenous adenine nucleotides. These requirements are consistent with conditions previously established to produce structural changes in mitochondria (Nicholls and Crompton, 1980). Moreover, the H^+-induced Ca^{2+} release is opposed by oligomycin and adenine nucleotides, which are reported to increase mitochondrial membrane stability (Harris, 1979). The ability of phospholipase activity inhibitors to prevent acid-induced Ca^{2+} release is in line with involvement of phospholipase in generalized permeability increases associated with low pH, Ca^{2+}, and phosphate (Palmer and Pfeiffer, 1981).

The generalized response of cardiac mitochondria in vitro to intracellular conditions mimicking ischemia (low pH, increased cytosolic Ca^{2+}, and rises in inorganic phosphate and fatty acid ester concentrations, as well as depleted matrix adenine nucleotides) is strongly suggestive that similar membrane changes may occur in vivo. This sequence of mitochondrial events would compromise the role of mitochondria in Ca^{2+} homeostasis and may contribute to the slow recovery of ATP levels (and contractile function) with reflow to the ischemic area.

VERAPAMIL PROTECTION

The goal of research into the mechanisms producing ischemic damage is to design rational interventions to prevent irreversible injury. Verapamil was introduced primarily as an antianginal agent and was included in a class of compounds able to inhibit the transmembrane Ca^{2+} flux in excitable tissues. In addition to its known role of reducing energy demand to improve ischemic recovery, verapamil has been shown to reduce the extent of damage to mitochondria during early stages of reversible injury. This effect may be an indirect consequence of decreasing intracellular Ca^{2+} levels (Henry et al., 1977). We have demonstrated protection of mitochondrial oxidative and Ca^{2+} transport function in hearts undergoing cardiopulmonary bypass with 60 minutes of ischemic arrest by a single bolus intracoronary infusion of verapamil in combination with KCl prior to the onset of ischemia (Pinsky et al., 1981). Verapamil has been shown by other investigators to have a protective effect on mitochondrial respiratory function (Clements et al., 1978) and to prevent phosphate-induced swelling of mitochondria in vitro (Vághy et al., 1981). Verapamil is also known to reduce ultrastructural damage both during coronary occlusion and following reperfusion with a more profound decrease in the extent of injury to mitochondria than to other organelles (Kloner et al., 1982).

Initial attempts to elucidate any interaction of Ca^{2+} antagonists with mitochondrial ATP production and Ca^{2+} uptake in vitro failed to describe any specific pharmacological effects on these processes. However, the ability of calcium antagonists to mimic the action of ruthenium red in inhibiting phosphate-induced swelling of heart mitochondria suggested that verapamil and diltiazem interact with mitochondrial Ca^{2+} homeostatic mechanisms (Vághy et al., 1981). Vághy and co-workers (1982) have demonstrated that drugs of this class are selective inhibitors of mitochondrial Na^+-induced Ca^{2+} release. Work on verapamil inhibition of Na^+-Ca^{2+} exchange activity was carried out in our laboratory on two preparations of canine heart mitochondria (Wolkowicz et al., 1983). Mitochondria isolated by polytron homogenization from ischemic heart in general appear more sensitive to ischemic damage than do mitochondria prepared by nagarse treatment (Pinsky et al., 1981; Watts, Koch, and LaNoue, 1980). Following loading of mitochondria with Ca^{2+} (30 ng-atoms/mg) and addition of ruthenium red to inhibit further energy-dependent Ca^{2+} uptake, Na^+-induced Ca^{2+} efflux was initiated by addition of 10 mM Na^+ (Fig. 6-2). Verapamil was found to have a marked inhibitory effect on Na^+-induced Ca^{2+} efflux in both preparations of mitochondria when phosphate is present (Fig. 6-2, Panel *A*, polytron homogenization; Panel *B*, nagarse treatment). The I_{50} for verapamil was 75% lower in the polytron preparation of mitochondria in the presence of phosphate than in a similar preparation in the absence of phosphate, as found by the Vághy group (1982). When acetate (20 mM) was substituted for phosphate (inset, Fig. 6-2) verapamil inhibition was significantly attenuated.

Further evidence for an anion dependence of the Na-Ca^{2+} exchange can be seen in the transformation of sigmoidal kinetics (Hill coefficient = 3) with respect to Na^+ concentration in the presence of phosphate, to hyperbolic kinetics with acetate as the compensating anion (Fig. 6-3, Panel *A*, polytron homogenization; Panel *B*, nagarse treatment). In Figure 6-3 the inset to

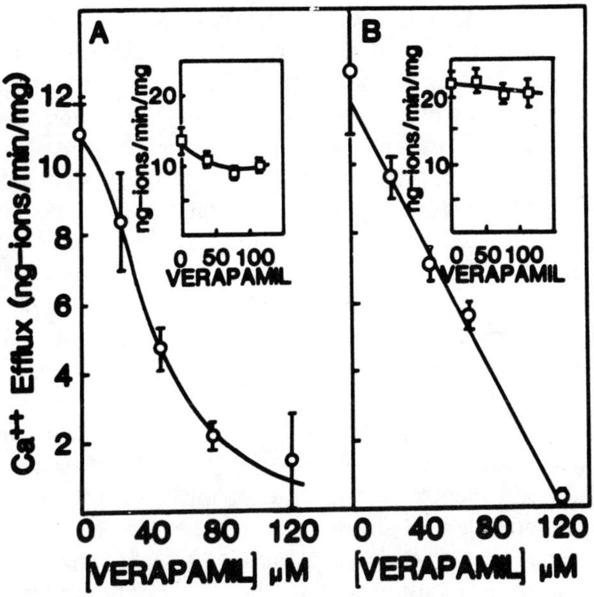

Figure 6-2. Inhibition of Na^{+}-Ca^{2+} exchange in canine heart mitochondria by verapamil in vitro: effect of anionic species (from Wolkowicz and McMillin-Wood, 1980*a*).

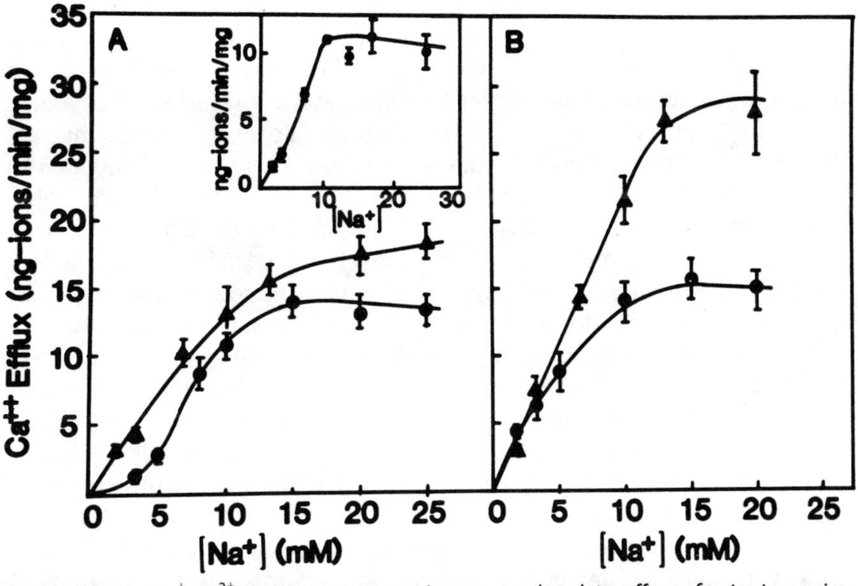

Figure 6-3. Na^{+}-Ca^{2+} exchange in canine heart mitochondria: effect of anionic species. The symbols refer to Na^{+}-induced Ca^{2+} efflux in the presence of 1.5 mM phosphate (● —— ●) and 20 mM acetate (▲ – ▲) (from Wolkowicz and McMillin-Wood, 1980*a*).

Panel *A* represents polytron mitochondria exposed to nagarse (25 μg/mg), and Na^+-induced efflux carried out in the presence of 1.5 mM phosphate. Both the V_{max} and the K_{Na^+} are significantly higher for both preparations of mitochondria when acetate is present. Crompton, Capano, and Carafoli (1976) were the first to report a sigmoidal dependence of Ca^{2+} efflux on Na^+ concentration in polytron-prepared heart mitochondria. Because no exogenous anion was added in these experiments (Crompton, Capano, and Carafoli, 1976) it is possible that endogenous phosphate associated with the mitochondria supported limited Ca^{2+} loading under these conditions.

Numerous references to phosphate and Ca^{2+}-induced permeability alterations in mitochondria exist in the literature (for review, see Nicholls and Crompton, 1980). In heart mitochondria, uncoupler or H^+-mediated Ca^{2+} loss is thought to take place by a combination of uniport reversal and a nonspecific pathway accompanied by loss of endogenous Ca^{2+} and adenine nucleotides (Jurkowitz and Brierley, 1982). These effects are believed to be distinct from electroneutral Na^+-Ca^{2+} exchange because the latter Ca^{2+} efflux occurs independent of any reduction in the mitochondrial membrane potential (Crompton, Capano, and Carafoli, 1976). However, Na^+-induced Ca^{2+} efflux from respiring rat heart mitochondria is inhibited by oligomycin and adenine nucleotides when phosphate is present in the incubation medium (Harris, 1979; Harris and Heffron, 1982). These studies suggest that Na^+-induced Ca^{2+} efflux may reflect an alteration in inner membrane permeability that is protected by binding of adenine nucleotides.

Under ischemic conditions, the ability of mitochondria to transport and retain Ca^{2+} is decreased as a function of occlusion time. Likewise, a time-dependent inhibition of the rates of Na^+-induced Ca^{2+} efflux in mitochondria occurs following coronary occlusion. Because verapamil appears to interact with the Na^+-Ca^{2+} exchange pathway in vitro, and because Ca^{2+} antagonists preserve mitochondrial ultrastructure and respiratory function during ischemia, experiments (Wolkowicz et al., 1983) were conducted to determine if any in vivo effects of verapamil on ischemic mitochondrial function could be measured. Verapamil was infused for 15 minutes prior to a 60 minute ligation of the left circumflex to produce myocardial ischemia. In the absence of verapamil, 1 hour of ischemia depressed Ca^{2+} uptake in heart mitochondria by 60% and decreased the maximum rate of Na^+-Ca^{2+} exchange activity in both preparations of mitochondria (Fig. 6-4). The sigmoidal relationship between Ca^{2+} efflux and Na^+ concentration observed in the polytron mitochondria became hyperbolic with ischemia. Pre-infusion of hearts with verapamil prior to 60 minutes of ischemia returned both the Na^+-Ca^{2+} exchange and Ca^{2+} uniport uptake toward normal values. In addition, treatment with the Ca^{2+} antagonist was associated with a return to sigmoidal kinetics (Hill coefficient = 2.5) for the polytron-prepared mitochondria.

Thus, it appears that verapamil infusion in dogs prior to 60 minutes of occlusion-induced ischemia results in significant maintenance of the normal rates and kinetics of not only Ca^{2+} uptake but also of Na^{+}-Ca^{2+} exchange activity of heart mitochondria. Preservation of Ca^{2+} homeostatic mechanisms during ischemia suggests a direct intracellular action of this drug, although indirect protective effects may also be a consequence of verapamil's role as a slow Ca^{2+} channel blocker. Watts, Koch, and LaNoue (1980) have concluded that verapamil improves cardiac functional recovery by decreasing energy demand during ischemia rather than by blocking Ca^{2+} influx during reperfusion, a conclusion based on a direct correlation between mechanical function and preservation of myocardial ATP stores due to either Ca^{2+} channel blocker action or KCl cardioplegia, which has been demonstrated to maintain the total adenine nucleotide pool.

In the present studies (Wolkowicz et al., 1983) the effect of verapamil to inhibit Na^{+}-induced Ca^{2+} efflux may correlate with its inhibitory action on mitochondrial swelling in the presence of phosphate or palmitoyl-CoA (Vághy et al., 1981). Thus, the protective action of verapamil during isch-

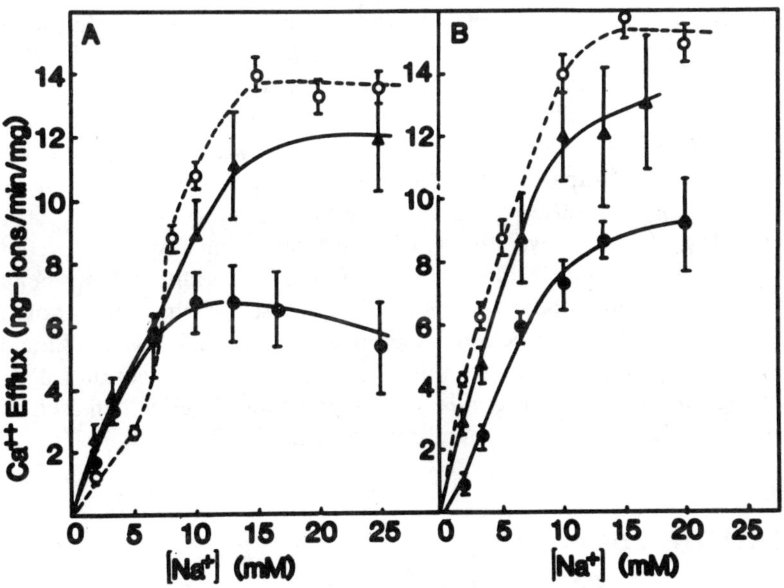

Figure 6-4. Effect of verapamil pretreatment of canine heart on Na^{+}-Ca^{2+} exchange activity following circumflex artery occlusion. The symbols represent rates of Na^{+}-induced Ca^{2+} efflux in control, nonischemic mitochondria (○ - - ○), in 60 minute ischemic mitochondria (● —— ●), and in 60 minute ischemic mitochondria following verapamil pretreatment of the heart (▲ — ▲).

emia on mitochondrial ultrastructure (Kloner et al., 1982) and Ca^{2+} homeostatic mechanisms may reflect interruption of a sequence of events producing osmotic changes with subsequent generation of a disrupted membrane structure. Prior studies on verapamil inhibition of Na^{+}-Ca^{2+} exchange, although carried out in the absence of phosphate, employed glutamate as respiratory substrate (Vághy et al., 1982). The sensitivity of NADH-linked respiration to Ca^{2+} inhibition reported by Chance (1965) may be an additional factor in the response of the Na^{+}-Ca^{2+} exchange process to Ca^{2+} antagonists.

CONCLUSIONS

Evidence suggests that mitochondria are involved in at least two phases of ischemic injury to the cardiac cell. During the period of reversible injury, diminished activity of Ca^{2+}-sequestering mechanisms at the sarcoplasmic reticulum and cell membrane, resulting from increased hydrogen ion production and decreased cellular ATP, will increase cytoplasmic Ca^{2+} to within a range necessary for mitochondrial uniport operation. Sequestration of cell Ca^{2+} can then take place at the expense of oxidative phosphorylation, which is limited by decreased oxygen delivery to the cell. This phase is followed by an apparent primary lesion in sarcolemmal membrane integrity that is expressed slowly during ischemia and leads to irreversible injury upon reperfusion. During this period, massive cellular influx of Ca^{2+} occurs with mitochondrial Ca^{2+} loading, deterioration of mitochondrial membrane integrity, and cell death. Verapamil and other drugs categorized as Ca^{2+} antagonists ameliorate the response of the cell to ischemic injury and may partially prevent the onset of irreversibility. Although verapamil may initially control Ca^{2+} influx into the cell by blocking the Ca^{2+} current, additional intracellular effects of the drug appear to be specific to preservation of mitochondrial integrity with related protection of oxidative and Ca^{2+} homeostatic function. It is presently unclear whether this action is a primary effect on Na^{+}-Ca^{2+} exchange activity or is secondary to a preventative effect on ischemic-induced alterations in inner membrane permeability.

ACKNOWLEDGMENTS

The authors wish to thank Dr. Lloyd H. Michael and Mr. Robert M. Lewis for their provision of physiological and surgical expertise in the ischemia studies. We are also grateful to Mrs. Corneille Smith for her typing of the manuscript. This work was supported by grants HL 23161 and HL 30186 from the National Institute of Health.

REFERENCES

Akerman, K. E. O., 1978, Effect of pH and Ca^{2+} on the retention of Ca^{2+} by rat liver mitochondria, *Arch. Biochem. Biophys.* **189**:256-262.

Asimakis, G. K., and L. A. Sordahl, 1977, Effects of atractyloside and palmitoyl coenzyme A on calcium transport in cardiac mitochondria, *Arch. Biochem. Biophys.* **179**:200-210.

Asimakis, G. K., and L. A. Sordahl, 1981, Intramitochondrial adenine nucleotides and energy-linked functions of heart mitochondria, *Am. J. Physiol.* **241**:H671-H678.

Beckman, J. K., K. Owens, T. E. Knauer, and W. B. Weglicki, 1982, Hydrolysis of sarcolemma by lysosomal lipases and inhibiton by chloropromazine, *Am. J. Physiol.* **242**:H652-H656.

Beinlich, C. J., M. G. Clark, E. E. McKee, J. A. Lins, and H. E. Morgan, 1981, Neutral-alkaline proteolytic activity in rat cardiac muscle cells, *J. Mol. Cell. Cardiol.* **13**:23-36.

Berne, R. M., and R. Rubio, 1974, Adenine nucleotide metabolism in the heart, *Circ. Res.* **23**:109-120.

Bersohn, M. M., K. D. Philipson, and J. Y. Fukushima, 1982, Sodium-calcium exchange and sarcolemmal enzymes in ischemic rabbit hearts, *Am. J. Physiol.* **242**:C288-C295.

Caroni, P., and E. Carafoli, 1981, The Ca^{2+}-pumping ATPase of heart sarcolemma. Characterization, calmodulin dependence and partial purification, *J. Biol. Chem.* **256**:3263-3270.

Chance, B., 1965, The energy-linked reaction of calcium with mitochondria, *J. Biol. Chem.* **240**:2729-2748.

Clements, I. P., R. E. Vlietstra, J. D. Dewey, and C. E. Harrison, 1978, Protective effect of verapamil infusion on mitochondrial respiratory function, in *Coronary Artery Disease*, M. Kaltenbach, P. Lichtles, R. Balcom and W. D. Bussmann, eds., Stuttgart, Thieme, pp. 284-295.

Corr, P. B., B. I. Lee, and B. E. Sobel, 1981, Electrophysiological and biochemical derangements in ischemic myocardium: Interactions involving the cell membrane, *Acta Med. Scand. Supp.* **651**:59-69.

Crompton, M., M. Capano, and E. Carafoli, 1976, The sodium-induced efflux of calcium from heart mitochondria, *Eur. J. Biochem.* **69**:453-462.

Fabiato, A., and F. Fabiato, 1978, Effects of pH on the myofilaments and the sarcoplasmic reticulum of skinned cells from cardiac and skeletal muscles, *J. Physiol* **276**:233-255.

Feher, J. J., F. N. Briggs, and M. L. Hess, 1980, Characterization of cardiac sarcoplasmic reticulum from ischemic myocardium: Comparison of isolated sarcoplasmic reticulum with unfractionated homogenates, *J. Mol. Cell. Cardiol.* **12**:427-432.

Fiskum, G., and R. S. Cockrell, 1978, Ruthenium red sensitive and insensitive calcium transport in rat liver and Ehrlich ascites tumor cell mitochondria, *FEBS Lett.* **92**:125-128.

Fleckenstein, A., 1977, Specific pharmacology of calcium in myocardium, cardiac pacemakers, and vascular smooth muscle, *Ann. Rev. Pharmacol. Toxicol.* **17**:149-166.

Fleckenstein, A., H. Tritthart, B. Fleckenstein, A. Herbst, and G. Grun, 1969, A new group of competitive Ca^{2+}-antagonists (iproveratril, D600, prenylamine) with highly potent inhibitory effects on excitation-contraction coupling in mammalian myocardium, *Pfluegers Arch.* **307**:R25.

Ganote, C. E., R. Seabra-Gomes, W. G. Nayler, and R. B. Jennings, 1975, Irreversible myocardial injury in anoxic perfused rat hearts, *Am. J. Pathol.* **80**:419-450.

Gross, R. W., R. C. Drisdel, and B. E. Sobel, 1983, Rabbit myocardial lysophospholipase-transacylase purification, characterization and inhibition by endogenous cardiac amphiphiles, *J. Biol. Chem.* **258**:15165-15172.

Hamm, C. W., and L. H. Opie, 1983, Protection of infarcting myocardium by slow channel inhibitors. Comparative effects of verapamil, nifedipine, and diltiazem in the coronary-ligated, isolated working rat heart, *Circ. Res.*, suppl. 1, **52**:129-138.

Harris, E. J., 1979, Modulation of Ca^{2+} efflux from heart mitochondria, *Biochem. J.* **178**:673-680.

Harris, E. J., and J. J. A. Heffron, 1982, The stimulation of the release of Ca^{2+} from mitochondria by sodium ions and its inhibition, *Arch. Biochem. Biophys.* **218**:531-539.

Hearse, D. J., 1977, Reperfusion of the ischemic myocardium, *J. Mol. Cell. Cardiol.* **9**:605-616.

Hearse, D. J., P. B. Garlick, and S. M. Humphrey, 1977, Ischemic contracture of the myocardium: Mechanisms and prevention, *Am. J. Cardiol.* **39**:986-993.

Henry, P. D., R. Schuchleib, J. Davis, E. S. Weiss, and B. E. Sobel, 1977, Myocardial contracture and accumulation of mitochondrial calcium in ischemic rabbit heart, *Am. J. Physiol.* **233**:H677-H684.

Idell-Wenger, J. A., L. W. Grotyohann, and J. R. Neely, 1978, Coenzyme A and carnitine distribution in normal and ischemic hearts, *J. Biol. Chem.* **253**:4310-4318.

Jurkowitz, M. S., and G. P. Brierley, 1982, H^+-dependent efflux of Ca^{2+} from heart mitochondria, *J. Bioenerg. Biomembr.* **14**:435-449.

Katz, A. M., and F. C. Messineo, 1981, Lipid-membrane interactions and the pathogenesis of ischemic damage in the myocardium, *Circ. Res.* **48**:1-16.

Kloner, R. A., L. W. V. DeBoer, N. Carlson, and E. Braunwald, 1982, The effect of verapamil on myocardial ultrastructure during and following release of coronary artery occlusion, *Exp. Mol. Pathol.* **36**:277-286.

Neely, J. R., M. J. Rovetto, J. T. Whitmer, and H. E. Morgan, 1973, Effects of ischemia on function and metabolism of the isolated working rat heart, *Am. J. Physiol.* **225**:651-658.

Nicholls, D. G., 1978, The regulation of extramitochondrial free calcium ion concentration by rat liver mitochondria, *Biochem. J.* **176**:463-474.

Nicholls, D. G., and M. Crompton, 1980, Mitochondrial calcium transport, *FEBS Lett.* **111**:261-268.

Palmer, J. W., and D. R. Pfeiffer, 1981, The control of Ca^{2+} release from heart mitochondria, *J. Biol. Chem.* **256**:6742-6750.

Pinsky, W. W., R. M. Lewis, J. B. McMillin-Wood, H. Hara, C. J. Hartley, P. C. Gillette, and M. L. Entman, 1981, Myocardial protection from ischemic arrest: Potassium and verapamil cardioplegia, *Am. J. Physiol.* **240**:H326-H335.

Pitts, B. J. R., C. A. Tate, W. B. Van Winkle, J. M. Wood, and M. L. Entman, 1978, Palmitylcarnitine inhibition of the calcium pump in cardiac sarcoplasmic reticulum: A possible role in myocardial ischemia, *Life Sci.* **23**:391-402.

Shen, A. C., and R. B. Jennings, 1972, Kinetics of calcium accumulation in acute myocardial ischemic injury, *Am. J. Pathol.* **67**:441-452.

Sordahl, L.A., and M. L. Stewart, 1980, Mechanisms of altered mitochondrial calcium transport in acutely ischemic canine hearts, *Circ. Res.* **47**:814-820.

Vághy, P. L., M. A. Matlib, L. Szekeres, and A. Schwartz, 1981, Protective effects of verapamil and diltiazem against inorganic phosphate induced impairment of oxidative phosphorylation of isolated heart mitochondria, *Biochem. Pharm.* **30**:2603-2610.

Vághy, P. L., J. D. Johnson, M. A. Mattlib, T. Wang, and A. Schwartz, 1982, Selective inhibition of Na^+-induced Ca^{2+} release from heart mitochondria by diltiazem and certain other Ca^{2+} antagonist drugs, *J. Biol. Chem.* **257**:6000-6002.

Watanabe, A. M., and H. R. Besch, Jr., 1974, Subcellular myocardial effects of verapamil and D600: Comparison with propranolol, *J. Pharmacol. Exp. Ther.* **191**:241-251.

Watts, J. A., C. D. Koch, and K. F. LaNoue, 1980, Effects of Ca^{2+} antagonism on energy metabolism: Ca^{2+} and heart function after ischemia, *Am. J. Physiol.* **238**:H909-H916.

Williamson, J. R., 1979, Mitochondrial function in the heart, *Ann. Rev. Physiol.* **41**:485-506.

Williamson, J. R., M. T. Woodrow, and A. Scarpa, 1975, Calcium binding to cardiac sarcolemma, in *Recent Advances in Studies on Cardiac Structure and Metabolism*, vol. 5, A. Fleckenstein and N. S. Dhalla, eds., University Park Press, Baltimore, pp. 61-71.

Wojtczak, L., 1976, Effect of long-chain fatty acids and acyl-CoA on mitochondrial permeability, transport and energy-coupling processes, *J. Bioenerg. Biomembr.* **8**:293-311.

Wolkowicz, P. E., and J. B. McMillin-Wood, 1980*a*, Respiration-dependent calcium ion uptake by two preparations of cardiac mitochondria, *Biochem. J.* **186**:257-266.

Wolkowicz, P. E., and J. B. McMillin-Wood, 1980*b*, Dissociation between mitochondrial calcium ion release and pyridine nucleotide oxidation, *J. Biol. Chem.* **255**:10348-10353.

Wolkowicz, P. E., and J. McMillin-Wood, 1981, Glutamate-supported calcium movements in rat liver mitochondria: Effect of anions and pH, *Arch. Biochem. Biophys.* **209**:408-422.

Wolkowicz, P. E., L. E. Michael, R. M. Lewis, and J. McMillin-Wood, 1983, Sodium-calcium exchange in dog heart mitochondria: Effects of ischemia and verapamil, *Am. J. Physiol.* **244**:H644-H651.

Wood, J. M., B. Bush, B. J. R. Pitts, and A. Schwartz, 1977, Inhibition of bovine heart Na^+, K^+-ATPase by palmitylcarnitine and palmityl-CoA, *Biochem. Biophys. Res. Commun.* **74**:677-684.

Wood, J. M., H. G. Hanley, M. L. Entman, C. J. Hartley, J. A. Swain, U. Busch, C-H. Chang, R. M. Lewis, W. J. Morgan, and A. Schwartz, 1979, Biochemical and morphological correlates of acute experimental myocardial ischemia in the dog, *Circ. Res.* **44**:52-61.

7

Mitochondria as a Target of Macrophage-Mediated Cell Injury

Donald L. Granger
Duke University Medical Center

John B. Hibbs, Jr.
University of Utah Medical Center

Albert L. Lehninger
Johns Hopkins University Medical Center

INTRODUCTION

There is evidence from in vivo and in vitro studies that mammalian host macrophages are capable of inhibiting division of, and in some cases killing, neoplastic cells (Hibbs, 1976). The cytotoxic effect of macrophages for neoplastic cells has been successfully studied in vitro using cell culture techniques. There are several important features of macrophage-mediated injury of neoplastic cells that have been demonstrated in a number of laboratories:

1. In order to injure neoplastic cells macrophages must undergo a differentiation process called activation (Hibbs et al., 1977). Activation occurs upon sequential signaling of macrophages by first, a secreted T lymphocyte product called gamma interferon, and second, exposure of macrophages to bacterial components such as lipopolysaccharide from gram negative organisms (endotoxin) (Weinberg, Chapman, and Hibbs, 1978; Meltzer, 1981).
2. Injury of neoplastic cells by activated macrophages requires direct contact between target and effector (Evans and Alexander, 1976), a phenomenon shown to involve cell-cell adhesion (Marino and Adams, 1980).

3. Activated macrophages injure neoplastic cells regardless of strain or species of origin of either the target or effector; the cytotoxic reaction is immunologically nonspecific (Fidler, 1978).
4. Macrophage-induced injury is manifested by cessation of target cell division. Whether target cell death follows depends on an as yet undefined property of the particular neoplastic cell line being studied (Cook, Hibbs, and Lewis, 1980).
5. Coincident with cell division arrest, mitochondrial O_2 uptake is inhibited in intact target cells as well as in permeabilized cells under state 3 conditions.

Under special conditions macrophages kill selected neoplastic cell types during a burst of O_2 consumption catalyzed by a membrane NADPH oxidase that produces superoxide. Superoxide then dismutates to dioxygen and hydrogen peroxide. There is good evidence that the H_2O_2 generated in this sequence kills the target cell (Nathan et al., 1979). For this reaction there is an absolute requirement for (1) a pharmacologic triggering stimulus such as phorbol myristate acetate (Nathan et al., 1979) or (2) antibody molecules absorbed to the surface of target cells (Nathan et al., 1980). Target cell death occurs rapidly within about 2 hours. However, most workers have studied macrophage cytotoxicity under conditions where triggering agents and antibodies are not present (Nathan, 1982). In these cases cytotoxicity runs a much more protracted course with target cell division ceasing at about 6 hours (Krahenbuhl and Remington, 1974), and cell death, when it occurs, takes 20 to 70 hours (Weinberg, Chapman, and Hibbs, 1978; Doe and Henson, 1978). It is noteworthy that not a single neoplastic cell type has been found resistant to macrophage-induced injury under these conditions (Hart and Fidler, 1981). The biochemical basis of macrophage-mediated injury of this latter type is poorly understood. Several macrophage products have been proposed as possible molecular mediators of neoplastic cell injury but there is no clear consensus on this point (Adams, 1982). How neoplastic cell division is halted and how cell death occurs is unknown. Furthermore, the question remains open as to which in vitro conditions may be most relevant to macrophage-target cell interactions within host tumor tissues.

ARREST OF L1210 LEUKEMIA CELL DIVISION IN VITRO BY MACROPHAGES

In general, studies on macrophage cytotoxicity have focused on death of neoplastic cells as a measure of cell injury. The transformed cell lines used as targets for macrophage killing have been selected on the basis of susceptibility to lysis rather than propensity to cause malignant disease. We, however, set out to find highly malignant cell types which might be resistant to macrophage-induced cytolysis. It was observed that lymphoblastic leukemia line 1210 (L1210) cells were not killed by cytotoxic macrophages (CM) under culture conditions where other transformed cell types were readily lysed (Hibbs and Granger, 1982). However, L1210 cells were clearly injured by CM because target cell division was totally arrested. This result of CM-induced injury (i.e., cytostasis not progressing to cytolysis) was observed by Krahenbuhl, Lambert, and Remington (1976) for a murine breast carcinoma cell line, EMT-6. The phenomenon was subsequently substantiated by Cook, Hibbs, and Lewis (1980), who surveyed a variety of cell lines transformed by different oncogenic viruses.

L1210 cells continue to divide for about 4 to 6 hours following initial exposure to CM; then target cell division stops (Fig. 7-1). During the ensuing 70 to 90 hours these cells fail to divide but remain viable. This result cannot

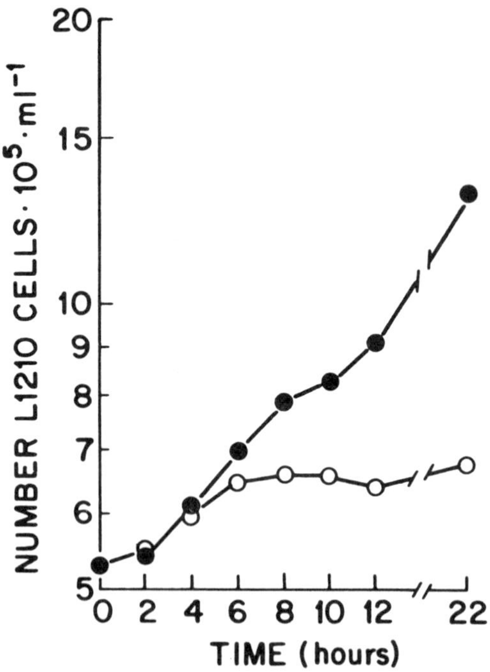

Figure 7-1. Induction of L1210 cell division arrest by CM. L1210 cells were cultured alone (●) and with CM (○). At the times shown cell counts were made. Viability remained greater than 95% in both types of cultures.

be due to the combined effects of division of a portion of target cells and death of another portion because the total number of L1210 cells remains constant while measurement of cell death by isotope release reveals that very few CM-injured L1210 cells die (Granger et al., 1980). After 20 hours coculture, L1210 cells were separated from CM, washed, and recultured in fresh medium. Despite removal from CM, injured L1210 cells failed to divide during a subsequent 50 hour culture period (Fig. 7-2). Once CM-induced cytostasis is established, the effect is independent of macrophages and persists long beyond the initial coculture period. Macrophage-induced cytostasis is not the result of depletion of a nutrient such as arginine (Currie, 1978), because L1210 cell division remains inhibited even with a supply of fresh nutrients (Fig. 7-2). Additionally, it was found that the medium from CM-L1210

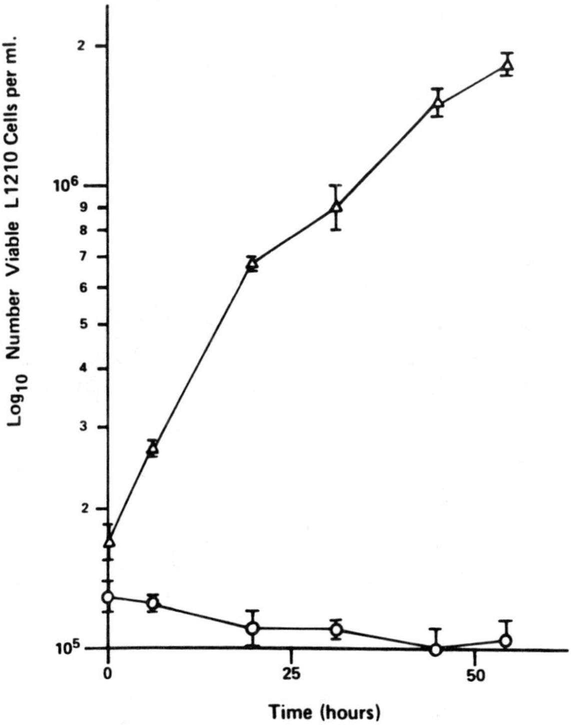

Figure 7-2. Separation of L1210 cells from macrophages does not reverse cytostasis. L1210 cells that had been exposed to CM (○) were separated from macrophages, washed by centrifugation, and recultured in fresh medium at Time (hours) = 0. △ = uninjured L1210 cells that were handled in exactly the same way except that they had not been exposed to CM.

cells cocultures supported exponential growth of uninjured L1210 cells (Fig. 7-3). These same arguments apply in ruling out the possibility that CM-induced cytostasis in this system results from the elaboration of a macrophage product into the culture medium that inhibits L1210 cell division (e.g., thymidine) (Stadecker et al., 1977).

The ultimate fate of CM-injured cells was studied. L1210 cells were cocultured with CM for 20 hours, then separated from macrophages, and recultured for prolonged periods with daily nutrient replenishment. At 60 to 80 hours following removal of CM, L1210 cell division recommenced with

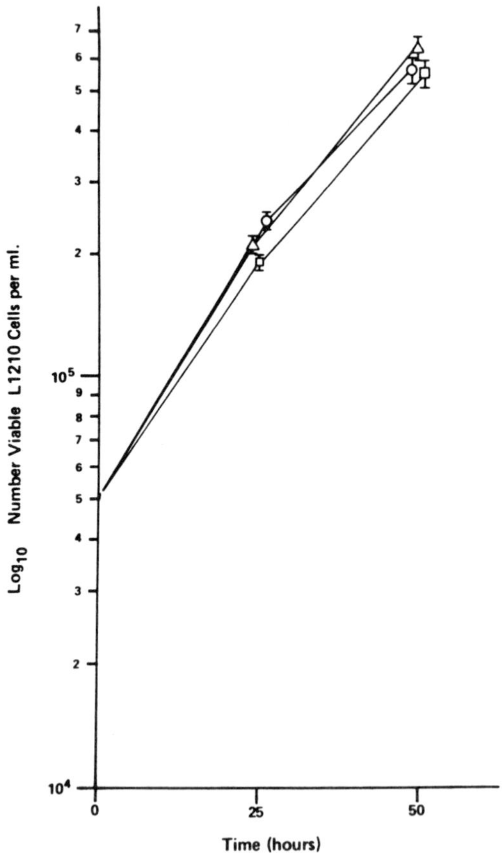

Figure 7-3. Medium from CM-L1210 cell cocultures supports multiplication of uninjured L1210 cells. Mediums for 20-hour cultures of CM alone (○) or CM + L1210 cells (□) were centrifuged to remove cells and then used to resuspend fresh L1210 cells for the growth curve measurements shown. △ = fresh culture medium.

the usual 12 to 13 hour generation time (Granger and Hibbs, 1981). Experiments using CM-injured L1210 cells cultured individually revealed that at least 50% of these cells recover their capacity for cell division. Macrophage-induced injury in this experimental system is a reversible process (Granger and Hibbs, 1981), a phenomenon also noted by Krahenbuhl, Lambert, and Remington in 1976.

INHIBITION OF NEOPLASTIC CELL RESPIRATION BY MACROPHAGES

It was found that L1210 cells injured by CM (i.e., by exposure to CM for 15-20 hours) developed an absolute requirement for a hexose. Nearly all CM-injured L1210 cells separated from macrophages and cultured in glucose-free medium died within 2 to 4 hours. In contrast, media containing D-glucose, D-mannose or D-fructose completely prevented injured L1210 cell death. Uninjured L1210 cells (log or stationary phase cells) or L1210 cells exposed to noncytotoxic macrophages remained totally viable and actually proliferated in medium without hexose. Inhibitors of glycolysis, such as 2-D-deoxyglucose, blocked the effect of glucose and led to rapid death of CM-injured L1210 cells at concentrations that had no effect on the viability of uninjured L1210 cells. Glycolysis rates were measured in CM-injured and uninjured L1210 cells and were found to differ markedly (Fig. 7-4). Uninjured L1210 cells had a clear Pasteur effect, that is, markedly decreased lactate production after addition of O_2. L1210 cells injured by CM exhibited higher aerobic glycolysis that was actually equal to the rate of anaerobic glycolysis, a result possible if oxidative phosphorylation was blocked in these cells. To test this possibility the rates of mitochondrial O_2 consumption were measured for both CM-injured and uninjured L1210 cells (Table 7-1). The rate of mitochondrial O_2 consumption in CM-injured L1210 cells was inhibited by nearly 90%. Therefore, it was unlikely that oxidative phosphorylation served as a significant source of chemical energy in these cells. Injured L1210 cells remained viable, producing chemical energy via glycolysis at a markedly increased rate, the physiologic response to inhibition of oxidative phosphorylation. These results explain why either lack of glycolytic substrate or inhibitors of glycolysis killed CM-injured L1210 cells.

LOCALIZATION OF THE TARGET CELL DEFECT TO THE MITOCHONDRIAL ELECTRON TRANSPORT CHAIN

Attempts were made to determine why L1210 cells injured by CM were unable to carry out oxidative phosphorylation. Isolation of mitochondria

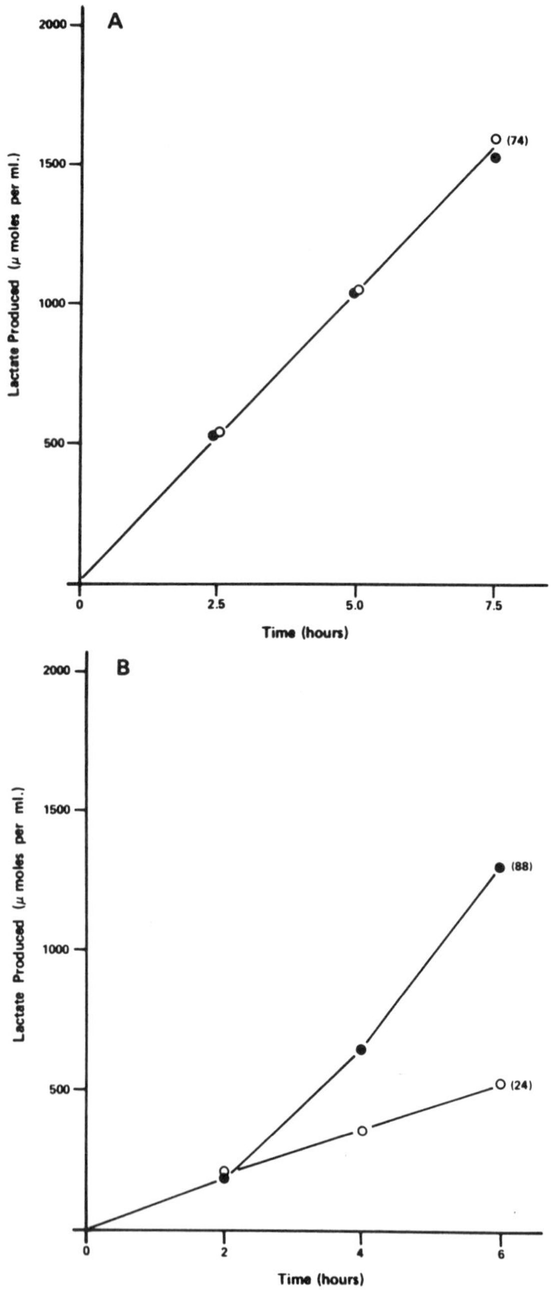

Figure 7-4. Glycolysis rates for CM-injured and uninjured L1210 cells. Aerobic (○) and anaerobic (●) glycolysis for L1210 cells that had been exposed to CM for 20 hours (*A*) and uninjured L1210 cells (*B*). Numbers in parentheses are maximal rates of lactate produced (nmoles per hour per 10^5 L1210 cells).

Table 7-1. Endogenous respiration of CM-injured L1210 cells

Number of Experiments	L1210 *Cells Cultured With*[a]	O_2 *Consumption*[b]
		μl/h · 10^6 cells
6	Alone	7.4 ± 0.3
4	Alone + ET (200 ng/ml)	7.2 ± 0.4
2	Alone + MAF (10% vol/vol)	7.2 ± 0.9
4	SM	6.9 ± 0.7
6	CM	1.1 ± 0.1
2	CM (MAF)	1.7 ± 0.4
3	Alone (oligomycin inhibited)[c]	0.9 ± 0.1
3	Alone (antimycin A inhibited)[c]	0.3 ± 0.1

[a]For these experiments, L1210 cells were cultured alone or with macrophages for 24 to 40 hours before respiration measurements.

[b]Values are the mean ± SEM for the number of experiments shown.

[c]The effects of oligomycin and antimycin A were determined by injection into the respiration vessel to a final concentration of 0.1 and 0.01 μM, respectively. Previous experiments showed that these concentrations produced maximal inhibition of uninjured L1210 cell respiration at 2×10^6 cells/ml. ET = endotoxin; MAF = macrophage-activating factor from sodium periodate-stimulated lymphocytes; SM = stimulated macrophages (peritoneal macrophages from mice given sterile proteose peptone ip 3 days before harvest); CM (MAF) = macrophages activated to the cytotoxic state by addition of MAF to the culture medium.

from CM-injured L1210 cells was unsuccessful because the number of target cells that could be injured using the standard in vitro assay system was too few for the procedures ordinarily used, but a method employing digitonin to selectively permeabilize the plasma membrane to mitochondrial substrates (Fiskum, 1980) proved to be useful. Experiments measuring rates of electron transport by following O_2 consumption showed that CM-injured L1210 cells had abnormal mitochondria (Granger and Lehninger, 1982). Oxidation of NAD-linked substrates and of succinate was reduced by 90% and 80%, respectively, compared to uninjured L1210 cells (Fig. 7-5). Assays for selected enzymes and segments of the electron transport chain showed that both complex I (NADH-coenzyme Q reductase) and complex II (succinate-coenzyme Q reductase) had markedly reduced activities (Granger and Lehninger, 1982), although electron transport from cytochrome c to O_2 occurred at a rate that was only slightly less than that obtained with mitochondria from uninjured L1210 cells (Fig. 7-5). Moreover, CM-injured L1210 cells that were made permeable to glycerophosphate oxidized this substrate at a rate indistinguishable from that of uninjured L1210 cells under state three conditions (Fig. 7-6). The respiratory control ratio (RCR) of injured L1210 cell respiration on glycerophosphate was also comparable to that of the uninjured cells, indicating that the oxidation of this substrate was coupled to the phosphorylation of ADP (Fig. 7-6). While oxidative phosphorylation, fueled by glycerophosphate, occurred under conditions where CM-injured L1210 cells were made permeable to this substrate present in the incubation medium, experiments with

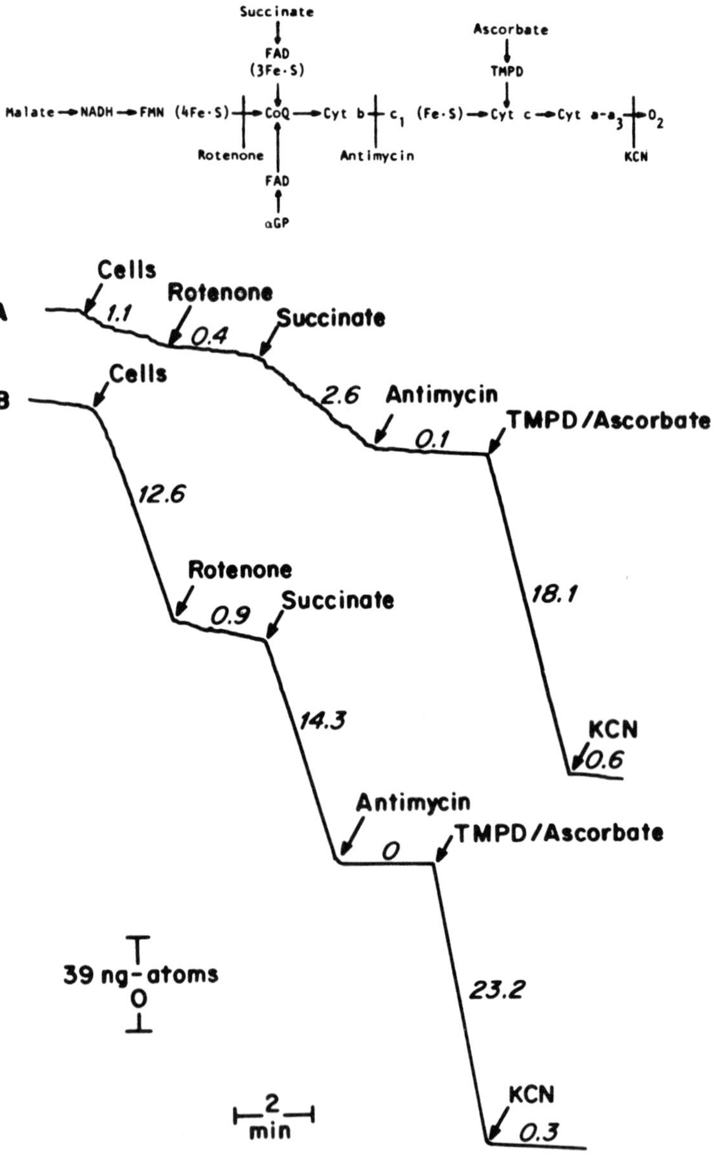

Figure 7-5. Addition of mitochondrial substrates to CM-injured L1210 cells permeabilized with digitonin. Polarographic traces of CM-injured cells (*A*) and uninjured cells (*B*) permeabilized with 0.005% digitonin in 0.25 M sucrose solution containing 5 mM K-malate, 2 mM KP_i, 10 mM $MgCl_2$, 2 mM ADP, and 10 mM K-HEPES, pH 7.1. Numbers are rates of O_2 consumption in nanogram atoms per minute per million cells. A scheme of the mitochondrial electron transport chain is given for reference.

intact L1210 cells under in vitro culture conditions revealed no evidence of either endogenous production or mitochondrial oxidation of glycerophosphate (Granger and Lehninger, 1982). This finding agreed with the previous finding that intact CM-injured L1210 cells were incapable of carrying out oxidative phosphorylation and hence relied totally on glycolysis for ATP production (Granger et al., 1980).

From these studies it was clear that CM-induced injury leads to selective defects in the respiratory chain of target L1210 cells, namely high level inhibition of complexes I and II. Because nearly all mitochondrial electron flow funnels through these enzyme complexes, CM-induced inhibition paralyzes oxidative phosphorylation. Time course experiments revealed that inhibition of complex I begins shortly before the arrest of target cell division,

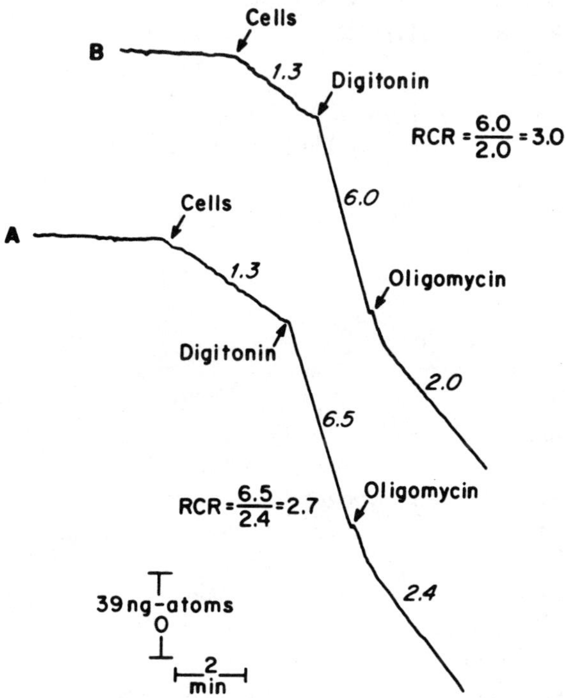

Figure 7-6. Oxidation of glycerophosphate by CM-injured L1210 cells permeabilized with digitonin. Polarographic traces of CM-injured L1210 cells (*A*) and uninjured L1210 cells (*B*) oxidizing glycerophosphate, 10 mM, in the medium described in Figure 7-5, but with 100 nM rotenone present. Oligomycin added (final concentration = 100 nM) to inhibit the mitochondrial ATPase. Respiratory control ratio (RCR) = state 3 rate (no oligomycin)/state 4 rate (with oligomycin).

leaving open the possibility that this metabolic defect may be related to the cytostatic event. In addition, CM-induced mitochondrial injury could lead to target cell death because some neoplastic cell types may be incapable of sustaining glycolysis at high rates required for energy-consuming processes vital to cell survival (e.g., ATP-driven ion translocations). Alternatively, intracellular acidification could occur within the nonrespiring CM-injured neoplastic cells, producing lactic acid at very high rates during long-term incubation (1-3 days). These results raise two critical questions: What do cytotoxic macrophages produce to cause these metabolic defects in neoplastic cells, and how does this CM product(s) cause shutdown of cell division and mitochondrial electron transport?

NONMITOCHONDRIAL OXYGEN CONSUMPTION IN CYTOTOXIC MACROPHAGE-INJURED L-1210 CELLS

During the course of experiments measuring mitochondrial O_2 uptake in intact L1210 cells, it was noted that CM-injured cells consumed O_2 at a low but constant rate in the presence of inhibitors that completely blocked mitochondrial electron transport, namely antimycin A or cyanide (Table 7-2). Log or stationary phase uninjured L1210 cells or L1210 cells incubated with noncytotoxic macrophages consumed O_2 at a much lower rate under these conditions (i.e., in the presence of cyanide). To avoid confusion, a short clarification may be useful. For intact L1210 cells growing under in vitro culture conditions, mitochondrial O_2 uptake accounts for 96% of all oxygen consumed (Granger et al., 1980). The residual nonmitochondrial O_2 uptake (measured with cyanide or antimycin A present) is very low, about 0.2 to 0.3 nanogram atoms per minute per million cells (Table 7-2). Macrophage injury reduced L1210 cell mitochondrial respiration by about 90% (Granger et al., 1980). Therefore, whole cell O_2 uptake is low because the majority of the oxygen consumed is mitochondrial. However, nonmitochondrial O_2 uptake (cyanide or antimycin A present) is increased 3- to 5-fold in L1210 cells injured by CM, compared to uninjured L1210 cells (Table 7-2). Thus, CM-induced injury results in inhibition of mitochondrial and induction of nonmitochondrial respiration in L1210 target cells. It is important to note that inhibition of mitochondrial respiration in and of itself is not sufficient to produce the cyanide-insensitive O_2 uptake observed for CM-injured L1210 cells. For example, inhibition of both complexes and I and II in uninjured L1210 cell mitochondria with rotenone and theonyltrifluoroacetone, respectively, did not lead to increased cyanide-insensitive O_2 consumption above the rate obtained in the presence of cyanide alone.

Table 7-2. Cyanide insensitive O_2 uptake in L1210 leukemia cells injured by cytotoxic macrophages

Incubation conditions prior to O_2 Uptake Measurements[a]	O_2 *Consumption Rate*[b] *(nanogram atoms O · min^{-1} · 10^{-6} cells)*
1. L1210 cells cocultured with CM for:	
4 hours	0.30
12 hours	1.46
20 hours	1.10 ± 0.09[c]
2. L1210 cells cocultured with CM for 20 hours; then separated from CM and contaminating CM removed by successive steps of adherence to plastic dishes[d]	1.02
3. CM cultured with L1210 for 20 hours; L1210 cells removed, discarded, and then CM detached with rubber policeman and CM O_2 consumption measured[e]	0.22
4. L1210 cells cocultured with noncytotoxic macrophages for 20 hours[f]	0.31
5. L1210 cells cultured alone:	
log phase cells	0.24 ± 0.02[c]
stationary phase cells	0.21

[a]L1210 cells were cultured together with macrophages or alone and then at the times given they were separated from macrophages by rinsing, washed by centrifugation, and resuspended in fresh culture medium for O_2 uptake measurements.

[b]Cells were suspended at approximately 5×10^6 per ml in culture medium (pH 7.4) containing 2.5 mM NaCN. Total volume was 1.3 ml in a continuously stirred chamber equipped with a Clark-type oxygen electrode. $T = 37°C$.

[c]Mean ± standard error of mean for eight separate experiments.

[d]L1210 cells removed from CM by rinsing were re-incubated in culture medium without serum in plastic dishes for 30 minutes. This procedure was repeated twice. The final L1210 cell suspension contained less than 1% contaminating cells from the CM monolayers.

[e]This procedure shows the rate of cyanide-insensitive O_2 uptake of approximately ten times the number of CM ordinarily present in the L1210 cell suspensions harvested by rinsing from CM monolayers.

[f]Peritoneal macrophages from mice injected three days previously with 1.0 ml sterile 10% proteose peptone.

When injured L1210 cells are removed from CM following the coculture incubation, some macrophages invariably dislodge from the culture vessel surface and contaminate the injured L1210 cell suspensions. However, the cyanide-insensitive O_2 uptake was not due to these contaminating macrophages, because after they were removed by successive stepwise adherence to plastic, there was no diminution of the rate of cyanide-insensitive respiration for the purified injured L1210 cells (Table 7-2). Furthermore, CM, detached with a rubber policeman, showed very little cyanide-insensitive respiration under these conditions (Table 7-2). The appearance of cyanide-insensitive O_2 uptake by injured L1210 cells correlated with the development of CM-induced cytostasis and inhibition of oxidative phosphorylation. The onset is between 4 and 12 hours after exposure to CM (Table 7-2, Fig. 7-1, and Granger and Lehninger, 1982). Using a method of continuous recording, the O_2 consumption with cyanide present continues at a constant rate for at least 3 hours in stirred CM-injured L1210 cell suspensions. The reaction requires that injured L1210 cells be intact because disruption by addition of detergent or by sonication completely stops cyanide-insensitive O_2 uptake. Inhibition also occurs if injured L1210 cells are permeabilized with 0.005% digitonin, suggesting that a cytosolic component is required for this reaction and may leak into and be diluted by the suspending medium. Inhibitors of cellular oxygenases, including azide, salicylhydroxamate, indomethacin, and nordihydroguaiacetate, had no effect.

CONCLUSIONS

The findings reported here provide an important avenue for investigating how macrophages injure neoplastic cells and may provide an explanation for the inhibition of target cell mitochondrial electron transport. They raise the possibility that the products of the cyanide-insensitive O_2 reduction observed in CM-injured L1210 cells are reactive oxygen intermediates such as superoxide (O_2^-), a hydroperoxide such as H_2O_2, or a product of these reactants, namely, hydroxyl radical. It is known that under special conditions macrophages can produce O_2^- in a burst of enzymatic activity due to activation of a membrane NADPH oxidase (Johnston, 1978). However, this reaction is short-lived, lasting about 2 hours, after which the phagocytes cannot be restimulated (Johnston, 1978). When CM injure neoplastic cells under the conditions described here, there is no stimulus to trigger phagocyte respiratory burst of O_2^- production. Moreover, as already discussed, the tempo of events occurs over a time span long beyond the minute-to-hour kinetics of the respiratory burst. For example, target cell division stops at 6 to 8 hours (Krahenbuhl and Remington, 1974); inhibition of electron transport occurs between 6 to 12 hours (Granger and Lehninger, 1982); target cell death, when it occurs, takes 18 to 70 hours, depending on the neoplastic cell line employed (Weinberg, Chapman, and Hibbs, 1978; Doe and Henson, 1978; Cook, Hibbs and Lewis, 1980).

It is possible that CM in some way initiate oxygen radical generation within target cells they contact. Once this occurs macrophage contact is no longer required. The target cells continue to generate toxic oxygen intermediates at a low but constant rate. It is pertinent that the cyanide-insensitive O_2 uptake of CM-injured L1210 cells proceeds at a constant rate for at least 3 hours, based on continuous recording with the oxygen electrode. A constant flux of one or more reactive O_2 intermediates could inhibit target cell DNA replication by reaction with nucleic acids (Lesko, Lorentzen, and Ts'o, 1980) or by oxidation of NADPH via the glutathione reductase/peroxidase cycle. Decrease in the NADP/NADPH ratio could interfere with ribonucleotide reduction and this inhibit DNA biosynthesis (Thelander and Reichard, 1979). It is also well known that O_2 radicals react with membrane unsaturated fatty acids to produce lipid hydroperoxides (McCay and Poyer, 1976). When lipid peroxidation occurs in the mitochondrial inner membrane, complexes I and II of the electron transport chain are inhibited (Narabayashi, Takeshige, and Minakami, 1982), discussed earlier as the loci at which electron transport is blocked in CM-injured L1210 cells (Granger and Lehninger, 1982).

Under a variety of conditions we have found that CM-induced injury of L1210 cells, as measured by cell division arrest, is always accompanied by inhibition of mitochondrial electron transport. The two phenomena thus far cannot be dissociated. These studies on target cell energy metabolism have provided potentially important knowledge for elucidating the mechanism of CM-induced cytostasis and cytolysis of tumor cells.

Cytostasis may be related to inhibition of mitochondrial metabolic flux because several mitochondrial metabolites are required for de novo nucleic acid synthesis.

Cytolysis may be related to inhibition of oxidative phosphorylation because some neoplastic cell types may be incapable of sustaining glycolysis at rates required for maintaining cellular ATP demands.

Any mechanism of CM-mediated neoplastic cell injury must account for the inhibition of mitochondrial electron transport because this inhibition appears to be a constant feature of this form of cytotoxicity.

ACKNOWLEDGMENTS

We thank Olive Sherman and Janet Routten for preparing the manuscript. This work was funded in part through Grant CA35893 from the National Institutes of Health and a grant from R. J. Reynolds Industries, Inc.

REFERENCES

Adams, D. O., 1982, Macrophage activation and secretion, *Fed. Proc.* **41**:2193–2197.

Cook, J. L., J. B. Hibbs, Jr., and A. M. Lewis, Jr., 1980, Resistance of simian virus 40-transformed hamster cells to the cytolytic effect of activated macrophages: A possible factor in species-specific viral oncogenicity, *Natl. Acad. Sci. (USA) Proc.* **77**:2886–2889.

Currie, G. A., 1978, Activated macrophages kill tumor cells by releasing arginase, *Nature* **273:**758-759.

Doe, W. F., and P. M. Henson, 1978, Macrophage stimulation by bacterial lipopolysaccharides. I. Cytolytic effect on tumor target cells, *J. Exp. Med.* **148:**544-556.

Evans, R., and P. Alexander, 1976, Mechanisms of extracellular killing nucleated mammalian cells by macrophages, in *Immunobiology of the Macrophage,* D. S. Nelson, ed., Academic Press, New York, pp. 535-576.

Fidler, I. J., 1978, Recognition and destruction of target cells by tumoricidal macrophages, *Israel J. Med. Sci.* **14:**177-191.

Fiskum, G., S. W. Craig, G. L. Decker, and A. L. Lehninger, 1980, The cytoskeleton of digitonin-treated rat hepatocytes, *Natl. Acad Sci. (USA) Proc.* **77:**3430-3434.

Granger, D. L., and J. B. Hibbs, Jr., 1981, Recovery from injury incurred by leukemia cells in contact with activated macrophages, *Fed. Proc.* **40:**761.

Granger, D. L., and A. L. Lehninger, 1982, Sites of inhibition of mitochondrial electron transport in macrophage-injured neoplastic cells, *J. Cell Biol.* **95:**527-535.

Granger, D. L., R. R. Taintor, J. L. Cook, and J. B. Hibbs, Jr., 1980, Injury of neoplastic cells by murine macrophages leads to inhibition of mitochondrial respiration, *J. Clin. Invest.* **65:**357-70.

Hart, I. R., and I. J. Fidler, 1981, The implications of tumor heterogeneity for studies on the biology and therapy of cancer metastasis, *Biochim. Biophys. Acta* **651:**37-50.

Hibbs, J. B., Jr., 1976, The macrophage as a tumoricidal effector cell: A review of *in vivo* and *in vitro* studies on the mechanism of the activated macrophage nonspecific cytotoxic reaction, in *The Macrophage in Neoplasia,* M. Fink, ed., Academic Press, New York, pp. 83-111.

Hibbs, J. B., Jr., and D. L. Granger, 1982, Activated macrophage-induced respiratory inhibition in neoplastic cells: Identification of lytic and nonlytic neoplastic cell phenotypes, in *Heterogeneity of macrophages,* O. Forster, ed., Academic Press, New York.

Hibbs, J. B., Jr., R. R. Taintor, H. A. Chapman, Jr., and J. B. Weinberg, 1977, Macrophage tumor killing: Influence of the local environment, *Science* **197:**279-282.

Johnston, R. B., Jr., 1978, Oxygen metabolism and the microbiocidal activity of macrophages, *Fed. Proc.* **37:**2759-2764.

Krahenbuhl, J. L., L. H. Lambert, and J. S. Remington, 1976, The effects of activated macrophages on tumor target cells: Escape from cytostasis, *Cell. Immunol.* **25:**279-93.

Krahenbuhl, J. L., and J. S. Remington, 1974, The role of activated macrophages in specific and nonspecific cytostasis and tumor cells, *J. Immunol.* **113:**507-516.

Lesko, S. A., R. J. Lorentzen, and P. O. P. Ts'o, 1980, Role of superoxide in deoxyribonucleic acid strand scission, *Biochemistry* **19:**3023-3028.

Marino, P. A., and D. O. Adams, 1980, Interaction of bacillus Calmette-Guerin-activated macrophages and neoplastic cells *in vitro.* II. The relationship of selective binding to cytolysis, *Cell. Immunol.* **54:**26-35.

McCay, P. B., and J. L. Poyer, 1976, Enzyme-generated free radicals as initiators of lipid peroxidation in biological membranes, in *The Enzymes of Biological Membranes: Electron Transport Systems and Receptors,* vol. 4, A. Martonosi, ed., Plenum Press, New York, pp. 239-256.

Meltzer, M. S., 1981, Macrophage activation for tumor cytotoxicity: Characterization of priming and trigger signals during lymphokine activation, *J. Immunol.* **127:**179-183.

Narabayashi, H., K. Takeshige, and S. Minakami, 1982, Alteration of inner-membrane components and damage to electron transfer activities of bovine submitochondrial particles induced by NADPH-dependent lipid peroxidation, *Biochem. J.* **202:**97-105.

Nathan, C. F., 1982, Secretion of oxygen intermediates: Role in effector functions of activated macrophages. *Fed. Proc.* **41:**2206-2211.

Nathan, C. F., L. H. Brukner, S. C. Silverstein, and Z. A. Cohn, 1979, Extracellular cytolysis by activated macrophages and granulocytes. I. Pharmacologic triggering of effector cells and the release of hydrogen peroxide, *J. Exp. Med.* **149:**84-99.

Nathan, C. F., L. Brukner, G. Kaplan, J. Unkeless, and Z. Cohn, 1980, Role of activated macrophages in antibody-dependent lysis of tumor cells, *J. Exp. Med.* **152:**183-197.

Stadecker, M. J., J. Calderon, M. J. Karnovsky, and E. R. Unanue, 1977, Synthesis and release of thymidine by macrophages, *J. Immunol.* **119:**1738-43.

Thelander, L., and P. Reichard, 1979, Reduction of ribonucleotides, *Ann. Rev. Biochem.* **48:**133-158.

Weinberg, J. B., H. A. Chapman, Jr., and J. B. Hibbs, Jr., 1978, Characterization of the effects of endotoxin on macrophage tumor cell killing, *J. Immunol.* **121:**72-80.

8

Mitochondrial Respiration and Calcium Transport in Tumor Cells

Gary Fiskum

George Washington University School of Medicine

INTRODUCTION

Mitochondria are involved in many activities that in one way or another affect the growth, development, and homeostasis of almost all eukaryotic cells. Two prime examples of these activities are the generation of ATP via oxidative phosphorylation and the regulation of the intracellular distribution of Ca^{2+} via mitochondrial Ca^{2+} influx and efflux.

In tumor cells a wide variety of cellular functions operate in an abnormal fashion, perhaps due to the abnormal expression of many differing genes or, more likely, to the altered expression of a relatively few genes that code for proteins whose functions influence the activities of numerous other enzymes or structural proteins. Thus, it is reasonable to pursue the possibility that alterations of certain key mitochondrial processes could be responsible for many of the cellular derangements that arise during the multistep process of cancer development. A comprehensive coverage of all the known mitochondrial abnormalities in tumor cells is beyond the scope of this report and can be found in a 1978 review by P. L. Pedersen. The objective of this review is to briefly summarize recent findings concerning the respiratory and Ca^{2+} transport characteristics of tumor cell mitochondria. The relationship of these characteristics to the survival-oriented phenotype and the chemotherapeutic sensitivity of tumor cells is also discussed.

RESPIRATION AND OXIDATIVE PHOSPHORYLATION

One of the earliest observations regarding the biochemistry of tumor cells was that they produce lactate via glycolysis at an abnormally high rate even in the presence of oxygen, which normally completely inhibits this process. Thus, tumor cells are often said to lack or exhibit a relatively weak Pasteur effect. Otto Warburg (1930, 1956) hypothesized that this characteristic was of central importance in the process of malignant transformation and that it was due to one or more defects in the capacity of tumor cell mitochondria to carry out oxidative phosphorylation. Subsequent investigations over the past 50 years have generally demonstrated a close relationship between maximal rates of glycolysis and the growth rate of malignant tumors (see Burk, Woods, and Hunter, 1967). However, some normal tissues (e.g., brain and retina) have also been shown to carry out aerobic glycolysis (Aisenberg, 1961). In addition, much evidence now indicates that glycolysis is often not the main source of cellular energy under physiological conditions (Reitzer, Wice, and Kennell, 1979; Stanisz, Wice, and Kennell, 1983). Nevertheless, many studies have shown that subtle alterations of mitochondrial respiration or respiration-linked activities could be important in explaining the abnormal energy metabolism, growth characteristics, and even drug sensitivity of many tumors.

Methods for Assaying the Energy-Linked Activities of Tumor Mitochondria

One of the main problems in assessing the normality of respiration and oxidative phosphorylation in tumor mitochondria has been methodological. Solid tumors, such as those of the Morris hepatoma series, contain a considerable amount of fat, connective tissue, and necrotic cells, resulting in an impairment of the functional integrity of the isolated mitochondria. Ascites tumor cells are also commonly used for mitochondrial studies because they are highly malignant, poorly differentiated, and easy to propagate and harvest. However, these, as well as most cell suspensions, are difficult to disrupt and effectively homogenize. Several methods have been developed to solve this problem (Wehrle and Pedersen, 1981), including the commonly employed Nagarse procedure, which involves the use of limited proteolytic digestion (Reynafarje and Lehninger, 1973; Pedersen et al., 1978; Kaschnitz, Hatefi, and Morris, 1976). However, each of these methods has been criticized on the basis that it often leads to variable degrees of mitochondrial damage (Pedersen, 1978; Scarpa and Graziotti, 1973), a criticism difficult to assess because there is little information concerning mitochondrial activities in situ.

In 1984 Moreadith and Fiskum described a new procedure for isolating ascites tumor mitochondria that uses low concentrations of digitonin to make the suspended cells susceptible to disruption by Teflon pestle/glass vessel homogenization. The functional integrity of these mitochondria, as reflected by their respiratory control ratios, was compared to those obtained by the Nagarse procedure and to that of nonhomogenized cells whose plasma membranes were selectively permeabilized with digitonin. Mitochondria isolated by the digitonin method from Ehrlich ascites tumor cells and two different lines of hepatoma cells exhibited normal to high rates of ADP-stimulated respiration and outstanding respiratory control ratios of 10.0 or more with NAD^+-linked substrates. These values were substantially higher than those obtained with mitochondria prepared by the Nagarse procedure (Pedersen et al., 1978) and those generally reported by investigators using a variety of techniques (Wehrle and Pedersen, 1981). It is particularly important that the respiratory control ratios of the isolated mitochondria were very similar to those obtained with suspensions of nonhomogenized, digitonin-permeabilized cells, as illustrated by the oxygen electrode tracings shown in Figure 8-1. The acceptor control ratio for 22aH mouse hepatoma mitochondria respiring on malate plus pyruvate was 225/23 = 9.8 (Fig. 8-1*A*) whereas that of digitonin-permeabilized 22aH hepatoma cells was 23/2.2 = 10.5 (Fig.

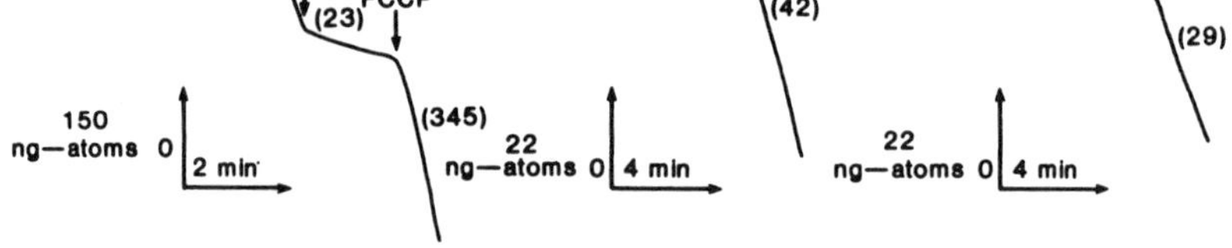

Figure 8-1. Oxygen electrode tracings of respiration by isolated hepatoma mitochondria and digitonin-permeabilized hepatoma cells. The conditions were as described by Moreadith and Fiskum (1984). The numbers in parentheses are the rates of oxygen consumption in ng-atoms O $min^{-1} \cdot mg^{-1}$ protein. FCCP = carbonylcyanide p-trifluoromethoxyphenyl-hydrazone. *A*: States 3, 4, and uncoupled respiration of isolated 22aH mouse hepatoma mitochondria. *B*: Digitonin titration of uncoupled 22aH tumor cell respiration in the presence of succinate and rotenone. *C*: States 3, 4, and uncoupled respiration of digitonin-permeabilized 22aH tumor cells (from Moreadith and Fiskum, 1984).

8-1*C*). Figure 8-1*B* describes the method used for titrating the suspended cells with digitonin (detailed in Moreadith and Fiskum, 1984).

Thus, a procedure now exists for measuring many normal and tumor mitochondrial activities in situ under simulated intracellular conditions and for isolating tumor mitochondria that apparently have undergone little if any damage during the isolation procedure. For the few lines of tumor cells that have been studied in this way, there is no evidence for significant differences between their respiratory control ratios and those of mitochondria from the corresponding normal tissue, such as liver. These findings are in contrast to previous reports (summarized in Pedersen, 1978), indicating that tumor mitochondria generally have lower maximal rates of respiration and lower respiratory control ratios than those of the control tissue.

Substrate Oxidation and Glycolysis

Irrespective of the method of isolation, many types of tumor mitochondria exhibit unusual selectivities toward the oxidizable substrates used as fuel for the electron transport chain, a fact particularly true for mitochondria from highly malignant hepatoma cells that, unlike normal liver mitochondria, do not appreciably oxidize fatty acids (Weinhouse, 1972). They must therefore rely on the oxidation of NADH derived from glycolysis and pyruvate or precursors of tricarboxylic acid cycle intermediates.

At one juncture it was suggested that the shuttle systems normally responsible for transferring reducing equivalents from cytosolic NADH into the mitochondrial respiratory chain were ineffective in tumor cells (Boxer and Devlin, 1961; Kovacevic, 1972). This scenario could explain the presence of significant aerobic lactate production in malignant tissues, based upon the need to regenerate the NAD^+ that is reduced during the glycolytic glyceraldehyde 3-phosphate dehydrogenase reaction. Thus, a buildup of NADH that cannot be effectively oxidized by the mitochondrial electron transport chain would stimulate the lactate dehydrogenase reaction that oxidizes NADH to NAD^+. Although the concept was attractive, it was soon disproved by Greenhouse and Lehninger (1976), who demonstrated the existence of an effective malate-aspartate NADH shuttle activity in six different lines of ascites tumor cells. An extension of these studies also demonstrated that approximately one third of the ATP generated in vivo by oxidative phosphorylation was due to the flow of electrons coming from cytosolic NADH via the mitochondrial malate-aspartate shuttle (Greenhouse and Lehninger, 1977).

Mitochondrial oxidation of cytosolic pyruvate has also been investigated as an explanation for the relatively weak Pasteur effect exhibited by tumor cells. Abnormally low rates of pyruvate oxidation have been reported for mitochondria isolated from tumors such as Ehrlich ascites cells (Paradies et al., 1983), erythroleukemias and chloromas (Abou-Khalil and Abou-Khalil, 1984), and Morris hepatomas (Paradies et al., 1983). However, it is also clear that mitochondria from some highly malignant tumors oxidize pyruvate in a normal fashion (particularly in the presence of malate (Moreadith and Fiskum, 1984)). In addition, erythroid and myeloid tumor mitochondria apparently oxidize pyruvate at abnormally low rates only at high pyruvate concentrations (>0.5 mM) but respire normally on pyruvate at lower, more physiological levels (0.1-0.5 mM) (Abou-Khalil and Abou-Khalil, 1984).

Several mechanisms have been proposed to account for the relatively poor oxidation of pyruvate by some tumor mitochondria. Studies where the kinetics of the purified pyruvate dehydrogenase complex from Ehrlich ascites tumor cells are compared with cellular metabolic flux measurements indicate that less than 15% of pyruvate produced from glycolysis can be oxidized by the Krebs cycle even when the enzyme is fully activated (Siess, Nimmannit, and Wieland, 1976; Lazo and Sols, 1980). An alternative explanation is that mitochondrial pyruvate oxidation is rate-limited by the activity of the inner-membrane pyruvate/OH^- transporter, a possibility supported by the correlation made by Paradies and co-workers (1983) between the rate of pyruvate-dependent respiration and the V_{max} of the pyruvate transporter for mitochondria from rat liver, Ehrlich ascites cells, and two types of hepatomas. More recently, Parlo and Coleman (1984) have shown that the low rate of pyruvate oxidation in Morris 3924A hepatoma mitochondria is at least partially due to the rapid efflux of pyruvate-derived mitochondrial citrate. This study also indicated that the abnormally elevated levels of cholesterol in tumor mitochondria may be responsible for the altered activity of the citrate (tricarboxylate) transporter and the consequential truncation of the Krebs cycle.

Another potential reason why much of the glycolytically produced pyruvate in tumor cells is converted into lactate rather than oxidized by the Krebs cycle is because there are generally not as many mitochondria in tumor cells as there are in corresponding normal cells (Pedersen, 1978), a situation particularly evident when normal hepatocytes are compared to rapidly proliferating hepatoma cells containing less than 50% the normal mitochondrial content, as indicated by morphometric electron microscopy studies (Howatson and Ham, 1955) and spectrophotometric determinations of mitochondrial cytochrome levels (Schrieber et al., 1970). These findings taken together with the often observed depression of pyruvate oxidation by isolated tumor mitochondria suggest that the maximum rate of pyruvate oxidation (and pyruvate-dependent oxidative phosphorylation) by tumor mitochondria in situ should be significantly below normal.

The maximum rate of energy-coupled respiration (state 3 respiration) is seldom reached in nonelectrically excitable cells except under extreme conditions where the cells are either recovering from an episode of de-energization (e.g., ischemia) or where cellular damage has resulted in an exceptionally high rate of ATP turnover. Although it appears that well oxygenated, mitochondria-deficient tumor cells normally have plenty of respiratory capacity to spare, many cells present within large, poorly vascularized tumors may vacillate between normoxic and hypoxic or ischemic conditions, thereby placing special demands on anaerobic and aerobic metabolic pathways. Thus, it is possible that the prevalence of glycolysis in tumor cells may actually represent a survival-oriented adaptation of malignant tissues.

Even if the cellular content of mitochondria or the activities of mitochondrial enzymes and transport proteins were to limit the utilization of glycolytic NADH and pyruvate, the metabolism of glucose to lactate would still be limited by the turnover of ATP and the availability of ADP and inorganic phosphate for glycolytic reactions involving substrate-level phosphorylation. This limitation has led some investigators to propose that tumor cells generate ADP and/or phosphate from ATP at an abnormally elevated rate, accounting at least partially for the presence of aerobic glycolysis and the high rates of anaerobic glycolysis that are typically observed with rapidly proliferating tumor cells. Evidence in favor of this hypothesis include findings that tumor cells possess both a highly active, mitochondrially bound hexokinase (Bustamante and Pedersen, 1977; Parry and Pedersen, 1983; Kurokawa et al., 1982) and one or more abnormally active ATP hydrolases, such as the Na^+/K^+ ATPase (Racker, 1972; Scholnick, Lang, and Racker, 1973; Racker, Johnson, and Blackwell, 1983).

The hexokinase reaction is particularly relevant to this problem because it is the first step of glycolysis, hydrolyzes ATP to form ADP and glucose-6-phosphate (but not P_i), and is subject to metabolic regulation. It is also significant that mitochondrially bound hexokinase may preferentially react with mitochondrially synthesized ATP (Gots, Gorin, and Bessman, 1972). The possibility that one or more highly active ATPases could be partially responsible for elevated glycolysis in tumor cells is also attractive in lieu of the fact that some of these reactions (e.g., the plasmalemmal Na^+/K^+ ATPase) utilize a relatively large proportion of the cellular ATP (Scholnick, Lang, and Racker, 1973) and generate both ADP and P_i. Although there are indications that the activity of either hexokinase (Kurokawa et al., 1982; Bustamante, Morris, and Pedersen, 1981) or the Na^+/K^+ ATPase (Racker, Johnson, and Blackwell, 1983; Suolina, Buchsbaum, and Racker, 1975) can limit the rate of glycolysis under specific conditions, no studies have convincingly demonstrated that this effect is any more pronounced in tumor cells than it is in the corresponding normal tissue. Nevertheless, it is a topic of considerable importance with regard to bioenergetics and metabolic regulation and should be further investigated.

Glutamine Oxidation

Substantial evidence indicates that glutamine is a major oxidizable mitochondrial substrate in tumor cells under physiologically realistic conditions (Reitzer, Wice, and Kennell, 1979; Stanisz, Wice, and Kennell, 1983; Lazo, 1981; Knox et al., 1967; Linder-Horowitz, Knox, and Morris, 1969; Knox, Horowitz, and Friedell, 1969; Kovacevic and Morris, 1972). In a 1981 study by Lazo, it was determined that 19% of steady-state ATP production by Ehrlich-Lettre ascites carcinoma cells is due to the oxidation of glutamine, compared to 1% for mitochondrial oxidation of glucose (via pyruvate). In addition, several different types of tumor mitochondria have been shown to rapidly oxidize glutamine at rates actually exceeding those for its deaminated product glutamate (Kovacevic and Morris, 1972; Abou-Khalil, Yunis, and Abou-Khalil, 1983; Kovacevic, 1971). These observations together with the finding of progression-linked expression of the deaminating enzyme glutaminase (Linder-Horowitz, Knox, and Morris, 1969; Knox, Horowitz, and Friedell, 1969) have led to an emerging picture of the catabolic reactions involved in the oxidation of glutamine by tumor cells (Fig. 8-2).

Cytosolic glutamine can rapidly cross the mitochondrial inner membrane (LaNoue and Schoolwerth, 1979) and be deaminated by glutaminase to form glutamate and ammonium ion (Kovacevic and Morris, 1972; Kovacevic, 1974; Moreadith and Lehninger, 1984*a*). Intramitochondrial glutamate primarily undergoes transamination with oxaloacetate via aspartate aminotransferase forming α-ketoglutarate and aspartate (Moreadith and Lehninger, 1984*a;* Eboli, Paradies, and Papa, 1976). Although glutamate can also be deaminated via glutamate dehydrogenase (particularly in some normal tissues, e.g., kidney) (Kovacevic, 1971; Schoolwerth, Nazar, and LaNoue, 1978) relatively little flux occurs through this route in tumor mitochondria (Moreadith and Lehninger, 1984*a*), possibly due to enzyme inhibition by NADPH or GTP.

In tumor cells and most normal tissues, the transamination reaction is limited by the production and availability of oxaloacetate. In normal mitochondria, the addition of malate promotes this reaction and subsequent oxidative reactions by forming oxaloacetate via malate dehydrogenase. However, with mitochondria isolated from five recently tested lines of tumor cells (including leukemias, thymomas, and hepatomas), the addition of malate stimulates glutamate oxidation but inhibits the production of aspartate (Moreadith and Lehninger, 1984*a*). This result was determined to be due to the activity of a progression-linked $NAD(P)^+$-dependent malic enzyme (Moreadith and Lehninger, 1984*a,* 1984*b;* Hansford and Lehninger, 1973; Sauer and Dauchy, 1978; Sauer et al., 1980) together with mitochondrial alanine aminotransferase (Moreadith and Lehninger, 1984*a*). Thus, in tumor

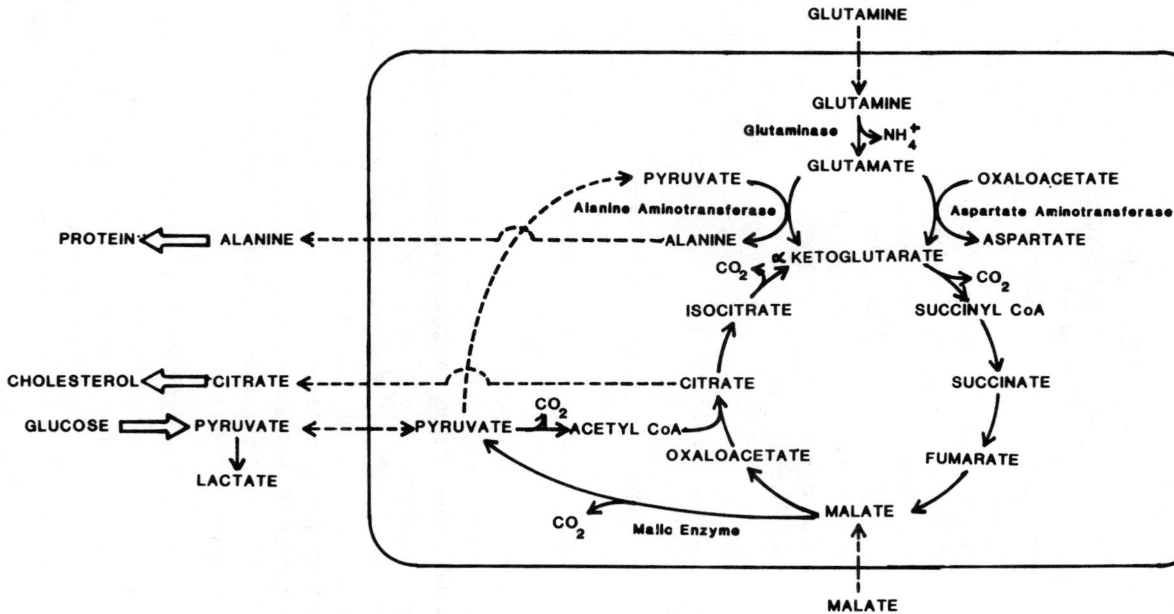

Figure 8-2. Pathways of glutamine and malate metabolism in tumor cells.

mitochondria, pyruvate can be formed directly from malate and transaminate with glutamate to form alanine and α-ketoglutarate (Fig. 8-2).

Glutamate-derived alanine is released from the mitochondria, whereas α-ketoglutarate is oxidized by the TCA cycle. In the presence of glutamate plus malate, the primary metabolite that is released from Ehrlich ascites tumor mitochondria in addition to alanine is citrate (Moreadith and Lehninger, 1984*a*), explainable by the citrate synthase reaction that requires acetyl-CoA in addition to the oxaloacetate generated from α-ketoglutarate. It appears that some of the pyruvate formed during the malic enzyme reaction is oxidatively decarboxylated by pyruvate dehydrogenase to form acetyl-CoA. The extent of citrate oxidation under these circumstances has yet to be determined but it may be severely limited due to the very rapid efflux of citrate from tumor mitochondria (Parlo and Coleman, 1984).

The following are important ramifications of this tumor-associated mode of glutamine metabolism:

1. The presence of intramitochondrial glutaminase allows the rapidly transported glutamine to be oxidized faster than when it is first converted to glutamate in the cytosol.
2. The presence of malic enzyme allows glutamine and other metabolites that can be converted into TCA cycle intermediates to be fully oxidized to CO_2 with little influx of other oxidizable substrates, for example, malate, and little or no oxidation of exogenous pyruvate.
3. Some of the pyruvate formed by the malic enzyme reaction may be released into the cytosol, or at least result in the inhibition of mitochondrial pyruvate uptake, thereby promoting lactate production and stimulating aerobic glycolysis.
4. NADPH produced by $NAD(P)^+$-dependent malic enzyme may be an important modulator of certain mitochondrial activities, for example, glutamate dehydrogenase or mitochondrial Ca^{2+} homeostasis.
5. The mitochondrial generation and release of alanine and citrate may play important roles in the accelerated anabolic pathways of protein and cholesterol synthesis in tumor cells.

Clearly, there is a significant potential impact of mitochondrial glutamine metabolism on the bioenergetics and growth characteristics of tumor cells. Further studies, including those performed with intact cells, will be necessary to define the physiological relationships between the oxidation of glutamine and the activity of other catabolic as well as an anabolic pathways.

Energy Coupling and Oxidative Phosphorylation

Under most physiological conditions, differences in the degree of coupling of mitochondrial respiration to ATP synthesis are much more important than differences in the maximal rate at which oxidative phosphorylation can occur. As mentioned earlier, although many reports have indicated that the respiratory control ratios of tumor mitochondria are generally lower than those of mitochondria from corresponding normal tissues, recent improvements in techniques used for isolating tumor mitochondria and measuring the energy-coupled respiration of tumor cells have questioned the significance of these findings. In addition, other types of energy-coupling determinations have indicated a striking similarity between mitochondria from normal tissues and mitochondria isolated from tumors by a variety of different methods. For instance, the ratio of ATP to oxygen consumed (P/O ratio) has generally been found to be normal for tumor mitochondria (i.e., the P/O is close to 3.0 for NAD^+-linked oxidizable substrates and close to 2.0 for FAD-linked substrates (Pedersen, 1978), a ratio not surprising in lieu of the fact that tumor mitochondria exhibit fair to good respiratory control and normal H^+ efflux stoichiometries for electron transport (Villalobo and Lehninger, 1979, 1980*a*). The importance of H^+ efflux stoichiometries pertains to the chemiosmotic mechanism of energy-coupling (Mitchell, 1966) stating that ATP synthesis is coupled to respiration via electron transport-dependent electrogenic H^+ efflux across the mitochondrial inner membrane and electrophoretic H^+ influx via the reversible H^+-pumping F_0F_1-ATPase.)

The intrinsically normal energy coupling of tumor mitochondria is reflected most convincingly by the level of ATP they can maintain in suspension under energized conditions. Although these data are limited, the mass action ratio for ATP synthesis ($[ATP]/[ADP][P_i]$) and the related phosphorylation potential have been determined for Ehrlich ascites tumor mitochondria (Villalobo and Lehninger, 1980*b*). Normal values were obtained in the presence of pyruvate plus malate, succinate, or ascorbate plus tetramethylphenylenediamine. These results taken together with results previously mentioned strongly suggest that the efficiency of energy coupling is normal at all three energy-conserving sites of the electron transport chain of Ehrlich and probably most other tumor mitochondria.

Although the coupling of respiration to oxidative phosphorylation by tumor mitochondria falls within the norm under standard in vitro conditions, ATP synthesis can be specifically inhibited by conditions relevant to the intracellular milieu of neoplastic tissues. Calcium, in particular, is one such factor that has often been shown to inhibit the synthesis of ATP by some normal and several different types of tumor mitochondria.

In the absence of ATP or Mg^{2+}, levels of accumulated intramitochondrial Ca^{2+} as low as 40 nmoles mg^{-1} mitochondrial protein have been reported to give half-maximal inhibition of ADP-stimulated oxygen uptake by Ehrlich or AS-30D hepatoma mitochondria (Villalobo and Lehninger, 1980*c*). More importantly, the abnormally high level of endogenous Ca^{2+} present within these mitochondria ($\sim$50 nmoles mg^{-1} protein) is apparently sufficient to significantly affect state 3 respiration (Villalobo and Lehninger, 1980*c;* Fiskum and Lehninger, 1981). Inhibition by Ca^{2+} has also been observed with digitonin-permeabilized tumor cells (Villalobo and Lehninger, 1980*c*), supporting the possible physiological significance of this effect. On the other hand, when cytosolic concentrations of ATP and Mg^{2+} are present, the inhibitory effects of Ca^{2+} on oxidative phosphorylation by Ehrlich (Villalobo and Lehninger, 1980*c*), hepatoma (Villalobo and Lehninger, 1980*c*), or myeloid tumor mitochondria (Abou-Khalil, Abou-Khalil, and Yunis, 1981) are partially or completed prevented. Nevertheless, the observation that tumor cells (Selkirk, Elwood, and Morris, 1971; Criss and Kakiuchi, 1982) and isolated tumor mitochondria (Eboli et al., 1979; Chan and Barbour, 1983; Lau and Chan, 1984) generally contain abnormally high levels of endogenous Ca^{2+} suggests that the influence of Ca^{2+} on mitochondrial and glycolytic ATP production by intact cells could be important.

These specific effects of Ca^{2+} on state 3 (ADP-stimulated) respiration are qualitatively different than the effects previously described for Ehrlich ascites and rat liver mitochondria (McIntyre and Bygrave, 1974*a;* Eboli, Cittadini, and Terranova, 1974) in which comparatively large levels of accumulated Ca^{2+} induced respiratory uncoupling as well as generalized respiratory inhibition. Inhibition of oxidative phosphorylation by Ca^{2+} under energy-coupled conditions is more likely to be due to inhibition of adenine nucleotide (ATP^{-4}/ADP^{-3}) translocation. This hypothesis was originally presented by Thorne and Bygrave over ten years ago (1974) and has received considerable experimental support (Eboli et al., 1979; Chan and Barbour, 1983; Lau and Chan, 1984; Moreno-Sanchez, 1983). However, abnormally low rates of adenine nucleotide transport by tumor mitochondria have also been ascribed to other factors such as low levels of intramitochondrial AMP (Chan and Barbour, 1983), elevated levels of membrane cholesterol (Lau et al., 1984), and an intrinsically low V_{max} for the adenine nucleotide translocator (Knowles, 1984).

Whatever the underlying mechanisms may be, the depressed rate of ATP/ADP exchange by tumor mitochondria may be important with regard to certain extremes of mitochondrial energy metabolism. On one hand, if an elevated rate of cellular ATP hydrolysis placed a demand for maximal ATP production, decreased adenine nucleotide transport could limit the rate of oxidative phosphorylation and place greater reliance on glycolytic, substrate-level phosphorylation. On the other hand, under anaerobic or ischemic conditions, depressed mitochondrial ATP influx and ADP efflux could limit the rate of mitochondrial ATP hydrolysis, which could be advantagous for the survival of the cell. The latter scenario might provide some physiological relevance to the well established observation that mitochondria from various tumors have a low uncoupler-stimulated ATPase activity (for examples see Knowles, 1982, 1984; Emmelot et al., 1959; LaNoue et al., 1974; Pedersen and Morris, 1974; Knowles and Kaplan, 1980; Hayashi, Gotoh, and Tagashira, 1980).

CALCIUM TRANSPORT

General Aspects of Mitochondrial Ca^{2+} Influx and Efflux

The transport of Ca^{2+} by mitochondria from various tissues has been studied extensively from both mechanistic and physiological viewpoints (for reviews see Fiskum and Lehninger, 1982; Fiskum, 1984; Akerman and Nicholls, 1983). Ca^{2+} uptake occurs via electrophoretic uniport in response to the positive outside (negative inside) membrane potential established by electron transport-dependent H^+ extrusion. Ca^{2+} efflux is a passive process that occurs via antiport with Na^+ in some types of mitochondria (e.g., those from heart and brain) or via another mechanism (possibly $Ca^{2+}/2H^+$ antiport) in other types of mitochondria (e.g., those from liver). Calcium influx is specifically inhibited by very low concentrations of ruthenium red, whereas no specific inhibitors of Ca^{2+} efflux have yet to be identified.

The V_{max} for Ca^{2+} uptake is at least 100 times greater than the V_{max} for efflux. This potential, put together with the fact that greater than 99% of intramitochondrial Ca^{2+} is present in a bound or precipitated form, provides energized mitochondria with the ability to sequester large amounts of exogenous Ca^{2+}. The capacity to accumulate Ca^{2+} is also greatly increased by the presence of intramitochondrial adenine nucleotides that apparently aid in precipitating Ca^{2+} and preventing Ca^{2+}-induced damage to the mitochondrial inner membrane.

Under steady-state conditions where Ca^{2+} efflux is saturated, the rates of influx and efflux are equivalent and very slow. The extramitochondrial free Ca^{2+} concentration that is maintained (or buffered) by mitochondrial Ca^{2+} influx-efflux cycling under these conditions falls within the range of approximately 0.5 to 1.5 μM. Since the basal cytosolic free Ca^{2+} concentration of most cells is now believed to be approximately 0.1 to 0.3 μM, the relatively high mitochondrial Ca^{2+} buffer point suggests that mitochondria are not primary regulators of cytosolic Ca^{2+}. However, work by Nicchitta and Williamson (1984) has shown that liver mitochondria can maintain the ambient free Ca^{2+} concentration at 0.25 μM when physiological concentrations of the polyamine spermine (0.5-1.0 mM) are present in the medium. Nevertheless, most lines of evidence indicate that the main roles of mitochondrial Ca^{2+} transport are to regulate the levels of intramitochondrial free Ca^{2+} and to protect the cell from large and prolonged elevations of cytosolic Ca^{2+}.

Capacities for Ca^{2+} Accumulation

Bygrave and co-workers (McIntyre and Bygrave, 1973, 1974*a;* Thorne and Bygrave, 1973) were the first to recognize that Ehrlich ascites tumor mitochondria can accumulate much more Ca^{2+} than rat liver mitochondria can without undergoing osmotic swelling and respiratory uncoupling. This observation subsequently has been confirmed by independent studies using both Ehrlich mitochondria (Villalobo and Lehninger, 1980*c;* Cockrell, 1981, 1982; Fiskum and Cockrell, 1985) and digitonin-permeabilized Ehrlich ascites tumor cells (Cockrell, 1981, 1982). Abnormally high Ca^{2+} uptake capacities have also been reported for leukemic lymphoblast mitochondria (Carpentieri and Sordahl, 1975) and AS-30D rat hepatoma mitochondria (Fiskum and Pease, 1984). The differences in the maximal capacity for Ca^{2+} accumulation between normal and tumor mitochondria are most pronounced in media containing NAD^{+}-linked oxidizable substrates and no ATP. However, significant differences are also observed in the presence of ATP (Table 8-1) or when mitochondria are respiring on succinate in the presence of rotenone.

The underlying reasons for the abnormally high Ca^{2+} uptake capacity of tumor mitochondria are unclear. Recent evidence suggests that the activity of a Ca^{2+}-activated mitochondrial phospholipase A2 may limit the net accumulation and retention of mitochondrial Ca^{2+} under many conditions (Beatrice, Palmer, and Pfeiffer, 1980; Fiskum et al., 1985). Moreover, Cockrell (1982) has shown that the phospholipase A2 inhibitor nupercaine elevates the maximal Ca^{2+} uptake of rat liver to a level comparable to that of Ehrlich tumor mitochondria but has no effect on the uptake of Ca^{2+} by digitonin-permeabilized Ehrlich cells or isolated Ehrlich tumor mitochondria. However, direct comparisons of mitochondrial phospholipase activities between normal and malignant tissues have yet to be reported.

It is highly unlikely that the maximum capacity for Ca^{2+} sequestration by mitochondria present within malignant or nonmalignant tissues is normally approached in vivo because in either case the physiological mitochondrial Ca^{2+} content is probably less than 1% of this value. The potential significance of the unusually high Ca^{2+} uptake of tumor mitochondria must therefore be related to pathological conditions, such as post-ischemic reperfusion (Fiskum, 1983), where energized mitochondria are exposed to abnormally high levels of intracellular Ca^{2+}. Under these conditions, the ability of mitochondria to accumulate enormous quantities of Ca^{2+} without becoming uncoupled may be essential in the maintenance of cell viability. Thus, as mentioned earlier with regard to high aerobic glycolysis and mitochondrial adenine nucleotide translocation, the alterations of Ca^{2+} uptake observed with tumor mitochondria may actually be an adaptation to the harsh environmental conditions to which malignant tissues are often exposed.

Table 8-1. Maximum Ca^{2+} uptake capacities of rat hepatoma and rat liver mitochondria[a]

Source	*−ATP*	*+ATP*
	nmoles Ca^{2+} accumulated / mg mitochondrial protein	
AS-30D Hepatoma cells ($n = 8$)	150 ± 25	1300 ± 200
Normal liver ($n = 7$)	30 ± 5	720 ± 85

[a]Calcium uptake was monitored with a Ca^{2+} electrode in a medium maintained at 30° containing 125 mM KCl, 5 mM HEPES (pH 7.0) 2 mM KH_2PO_4, 1 mM $MgCl_2$, 5 mM glutamate and 5 mM malate in the absence or presence of 3 mM Na_2 ATP plus 3 mM $MgCl_2$.

Steady-State Buffering of Ambient Free Ca^{2+}

Most investigators of mitochondrial Ca^{2+} transport currently believe that under physiological conditions the main function of mitochondrial Ca^{2+} influx and efflux is to regulate the concentration of intramitochondrial free Ca^{2+} (Denton and McCormack, 1980). This function could be very important to cellular energy metabolism because several key mitochondrial enzymes (e.g., pyruvate and α-ketoglutarate dehydrogenases) are regulated by changes in the matrix Ca^{2+} concentration (Denton and McCormack, 1980). It is, however, quite possible that mitochondrial Ca^{2+} transport could assist in the regulation of both intramitochondrial and extramitochondrial Ca^{2+} levels, a possibility supported by the finding that spermine can lower the mitochondrial Ca^{2+} buffer point to a level similar to the cytosolic free Ca^{2+} concentration (Nicchitta and Williamson, 1984). It is also largely unknown what the effects of mitochondrial Ca^{2+} buffering are on the transient elevations of cytosolic Ca^{2+} induced by hormonal, electrical, and mitogenic stimulation of Ca^{2+} influx across the plasmalemma or Ca^{2+} efflux from the endoplasmic (or sarcoplasmic) reticulum.

If it is assumed that cytosolic Ca^{2+} levels can rise to a point where endoplasmic reticulum Ca^{2+} accumulation becomes saturated or where net mitochondrial Ca^{2+} uptake is elicited, then the relative Ca^{2+} buffer points of normal versus tumor mitochondria would be physiologically important. In this regard, preliminary results suggest that tumor mitochondria buffer the ambient free Ca^{2+} concentration of cytosol-like media at higher levels than those maintained by normal mitochondria (Fiskum, 1981; Fiskum, Becker, and Lehninger, 1980). However, rigorous comparisons of Ca^{2+} buffering by mitochondria from normal and corresponding malignant tissues have not been reported. Our experiments with both isolated rat AS-30D hepatoma mitochondria and digitonin-permeabilized AS-30D cells indicate that their extramitochondrial Ca^{2+} buffer point is 1.0 to 1.5 μM in the presence of physiological concentrations of K^+, Mg^{2+}, P_i and ATP (Fiskum, 1985). Under the same conditions, rat liver mitochondria buffer the ambient free Ca^{2+} at 0.5 to 0.8 μM (Nicchitta and Williamson, 1984; Becker, Fiskum, and Lehninger, 1980). Considerably larger differences are observed in media containing Na^+. In the presence of 20 mM Na^+, the buffer points of AS-30D and rat liver mitochondria are 2.0 to 2.5 μM and 0.8 to 1.0 μM respectively (G. Fiskum, unpublished results). Significant elevation of the Ca^{2+} buffer point by Na^+ is also observed with suspensions of Ehrlich ascites tumor mitochondria (Fleschner et al., 1983) and rat (NEDH) insulinoma mitochondria (Prentki, Janjic, and Wollheim, 1983). In all cases, the presence of 20 mM Na^+ causes an increase in the Ca^{2+} buffer point of at least 0.5 μM. The effect of Na^+ on mitochondrial Ca^{2+} buffering can be attributed to an acceleration of ruthenium red-insensitive Ca^{2+} efflux that in Na^+-responsive mitochondria is apparently mediated by a $Ca^{2+}/2Na^+$ antiporter.

Results from these few physiological studies on mitochondrial Ca^{2+} buffering suggest that the cytosolic free Ca^{2+} concentration of tumor cells could reach abnormally high levels under conditions where mitochondrial Ca^{2+} transport becomes involved. Although there is currently no direct evidence to support this hypothesis, it is consistent with the observation that tumor cells often possess unusually high levels of total Ca^{2+} (Selkirk, Elwood, and Morris, 1971; Criss and Kakiuchi, 1982). The accelerated turnover of phosphatidylinositol that is associated with oncogene-induced transformation might also be expected to elevate cytosolic Ca^{2+} (Macara, 1985), an elevation presumably due to the production of inositol trisphosphate, which is known to trigger the release of Ca^{2+} from the endoplasmic reticulum (Williamson et al., 1985). It is proposed that the extent to which cytosolic Ca^{2+} becomes elevated could be limited by mitochondrial Ca^{2+} buffering. Thus, it is possible that a connection exists between the unusual Ca^{2+} transport properties of tumor mitochondria and the basic neoplastic phenotype of tumor cells. Experimental support for this hypothesis will await further information concerning the distribution of intracellular Ca^{2+} together with its influence on the structure, metabolism, and growth characteristics of normal and malignant tissues.

SUMMARY AND PROSPECTIVE

Results obtained from recent studies using improved methods for isolating mitochondria from ascites tumor cells and for measuring mitochondrial oxygen consumption within permeabilized cells have strengthened the conclusion that there is no generalized deficiency in the rate of respiration or the energy coupling of tumor mitochondria. Although the maximum rate at which some substrates are oxidized may be relatively slow, the oxidation of other substrates (e.g., glutamine) can be unusually rapid. Several lines of evidence also indicate that altered mitochondrial ATP production is not a viable explanation for the lack of the Pasteur effect in tumor cells. However, the abnormal metabolism of pyruvate, glutamine, and other substrates may result in alterations of regulatory factors that could promote accelerated cholesterol synthesis as well as aerobic glycolysis. Further metabolic studies performed under physiological conditions with intact normal and malignant cells will be necessary to determine what combination of enzyme and/or transport activities is actually responsible for altered energy metabolism in tumor cells.

Several different types of tumor mitochondria exhibit an unusually large capacity for accumulating and retaining exogenous Ca^{2+}. It is postulated that this characteristic, together with other abnormalities, provides malignant tissues with the ability to experience fluctuations in the supply of oxygen and oxidizable substrates without undergoing irreversible cellular injury due to Ca^{2+}-induced damage to the mitochondria and other cellular membranes. The physiological buffering of intramitochondrial and extramitochondrial Ca^{2+} by mitochondrial Ca^{2+} influx and efflux is another topic of interest among investigators of cellular Ca^{2+} homeostasis. Preliminary results indicate that mitochondria from some lines of tumor cells have a higher extramitochondrial Ca^{2+} buffer point than do mitochondria from the corresponding normal tissue. This finding may relate in turn to other aspects of Ca^{2+} metabolism in tumor cells, involving oncogene-associated alterations in the metabolism of specific membrane phospholipids.

The underlying mechanisms responsible for the abnormal transport of Ca^{2+}, adenine nucleotides, and oxidizable substrates by tumor mitochondria remain obscure. However, several lines of evidence suggest that a substantially elevated level of membrane cholesterol could be at least partially responsible for some of these properties.

The information obtained from studies on tumor mitochondria has not, as yet, resulted in the development of any new forms of therapy for cancer. Several antitumor drugs do, however, significantly influence mitochondrial activities and in some cases exhibit specificity toward tumor mitochondria. Lonidamine (Floridi and Lehninger, 1983), rhodamine 123 (Modica-Napolitano et al., 1984), adriamycin (Muhammed, Ramasarma, and Kurup, 1982), cisplatinum (Binet and Volfin, 1977), and methylglyoxal-bis(guanylhydrazone) (Pleshkewych et al., 1980) all either inhibit or uncouple mitochondrial respiration. It is unclear whether the antiproliferative actions of these agents are directly related to their inhibition of oxidative phosphorylation. Nevertheless, there are indications that respiratory inhibition or uncoupling by adriamycin and methylglyoxal-bis(guanylhydrazone) is partially responsible for the profound toxic side effects these drugs have on the heart (Pollakis, Goormaghtigh, and Ruysschaert, 1983) and the gastrointestinal tract (Pleshkewych, Maurer, and Porter, 1983).

The antitumor agents lonidamine and rhodamine 123 are particularly interesting because they appear to possess some specificity toward tumor mitochondria. Although lonidamine inhibits respiration by both normal and neoplastic tissues, it stimulates aerobic glycolysis in normal cells and inhibits the lactate production of tumor cells (Floridi et al, 1981). Inhibition of aerobic glycolysis may be closely related to the cytostatic activity of this drug and correlates with a specific inhibition of mitochondrially bound hexokinase (Floridi et al., 1981). The tumor-associated presence of this form of hexokinase could therefore explain the relatively specific antitumor effects of lonidamine.

Rhodamine 123 is a fluorescent dye that can be used to selectively stain mitochondria within living cells. Its mitochondrial specificity is apparently due to the large, negative-inside, mitochondrial membrane potential that causes this positively charged, amphipathic molecule to be drawn into the matrix. Rhodamine 123 also has a wide variety of deleterious effects on mitochondria isolated from normal and neoplastic cells. These range from the direct inhibition of the electron transport chain and the F_0F_1 ATP synthetase (Modica-Napolitano et al., 1984) to the inhibition of the transport and processing of cytosol-synthesized precursors of mitochondrial matrix enzymes (Morita, 1982). These effects may be responsible for the cytotoxicity of rhodamine 123, which at low concentrations is relatively specific for carcinoma cells (Lampidis et al., 1983). Carcinoma specificity appears to be due to an unusually long (3-5 days) retention time for rhodamine by this class of tumor cells (Summerhayes et al., 1982). Preliminary evidence indicates that the rhodamine retention time is dependent on the magnitude of the mitochondrial membrane potential. Thus, carcinoma mitochondria may maintain a relatively large membrane potential compared to other normal and neoplastic cells that are relatively resistant to the cytotoxic effects of this drug. This possibility, however, remains to be confirmed by independent methods.

ACKNOWLEDGMENTS

This review article was written while the author was supported by grant CA32946 from the National Institutes of Health. Part of the work described herein was supported by the same grant.

REFERENCES

Aisenberg, A. C., 1961, *The Glycolysis and Respiration of Tumors,* Academic Press, New York.

Akerman, K. E. O., and D. G. Nicholls, 1983, Physiological and bioenergetic aspects of mitochondrial calcium transport, *Rev. Physiol. Biochem. Pharmacol.* **95:**149-201.

Abou-Khalil, S., W. H. Abou-Khalil, and A. A. Yunis, 1981, Inhibition by Ca^{2+} of oxidative phosphorylation in myeloid tumor mitochondria, *Arch. Biochem. Biophys.* **209:**460-464.

Abou-Khalil, S., and W. H. Abou-Khalil, 1984, Pyruvate oxidation by erythroleukemia and chloroma tumor mitochondria, *Fed. Proc.* **43:**1978.

Abou-Khalil, W. H., A. A. Yunis and S. Abou-Khalil, 1983, Prominent glutamine oxidation activity in mitochondria of hematopoietic tumors, *Cancer Res.* **43:**1990-1993.

Beatrice, M. C., J. W. Palmer, and D. R. Pfeiffer, 1980, The relationship between mitochondrial membrane permeability, membrane potential and the retention of Ca^{2+} by mitochondria, *J. Biol. Chem.* **255:**8663-8671.

Becker, G., G. Fiskum, and A. L. Lehninger, 1980, Regulation of free Ca^{2+} by mitochondria, endoplasmic reticulum and digitonin-treated cells, *J. Biol. Chem.* **255:**9009-9012.

Binet, A., and P. Volfin, 1977, Effect of an anti-tumor platinum complex, Pt (II) diamino-toluene, on mitochondrial membrane properties, *Biochim. Biophys. Acta* **461:**182-187.

Boxer, G. E., and T. M. Devlin, 1961, Pathways of intracellular hydrogen transport, *Science* **134:**1495-1497.

Burk, D., M. Woods, and J. Hunter, 1967, On the significance of glucolysis for cancer growth, with special reference to Morris hepatomas, *J. Natl. Cancer Inst.* **38:**839-863.

Bustamante, E., H. P. Morris, and P. L. Pedersen, 1981, Energy metabolism of tumor cells: Requirement for a form of hexokinase with a propensity for mitochondrial binding, *J. Biol. Chem.* **256:**8699-8704.

Bustamante, E., and P. L. Pedersen, 1977, High aerobic glycolysis in rat hepatoma cells in culture: Role of mitochondrial hexokinase, *Natl. Acad. Sci. (USA) Proc.* **74:**3735-3739.

Carpentieri, U., and L. A. Sordahl, 1975, Functional properties of mitochondria isolated from murine L5178 lymphoblasts grown in cell culture, *Cancer Res.* **35:**3636-3640.

Chan, S. H. P., and R. L. Barbour, 1983, Adenine nucleotide transport in hepatoma mitochondria: Characterization of factors influencing the kinetics of ADP and ATP uptake, *Biochim. Biophys. Acta* **723:**104-113.

Cockrell, R. S., 1981, Calcium transport by Ehrlich ascites tumor cell mitochondria *in vitro* and *in situ, Arch. Biochem. Biophys.* **212:**443-451.

Cockrell, R. S., 1982, The influence of nupercaine on Ca^{2+} transport by rat liver and Ehrlich ascites cell mitochondria, *FEBS Lett.* **144:**279-282.

Criss, W. E., and S. Kakiuchi, 1982, Calcium: Calmodulin and cancer, *Fed. Proc.* **41:**2289-2291.

Denton, R. M., and J. G. M. McCormack, 1980, On the role of the calcium transport cycle in heart and other mammalian mitochondria, *FEBS Lett.* **119:**1-8.

Eboli, M. L., A Cittadini, and T. Terranova, 1974, Effect of low calcium concentration on the oxidation of NAD-linked substrates in rat liver and tumor mitochondria, *Experientia* **30:**1125-1127.

Eboli, M. L., G. Paradies, and S. Papa, 1976, Transport of anionic substrates and glutamate metabolism in mitochondria from ascites tumor cells, *Cancer Res.* **36:**3119-3125.

Eboli, M. L., K. Malmstrom, T. Galeotti, L. Lopez-Alarcon, and E. Carafoli, 1979, Calcium transport and translocation of adenine nucleotides in mitochondria from Morris hepatoma 3924A, *Cancer Res.* **39:**2737-2742.

Emmelot, P., C. J. Bos, P. J. Brombacher, and J. F. Hampe, 1959, Studies on isolated tumor

mitochondria: Biochemical properties of mitochondria from hepatomas with special reference to a transplanted rat hepatoma of the solid type, *Br. J. Cancer Res.* **13:**348-379.
Fiskum, G., 1981, Mitochondrial respiration and Ca^{2+} transport measured in digitonin-permeabilized tumor cells, *Fed. Proc.* **40:**1563.
Fiskum, G., 1983, Involvement of mitochondria in ischemic cell injury and intracellular calcium, *Am. J. Emerg. Med.* **1:**147-153.
Fiskum, G., 1984, Physiological aspects of mitochondrial calcium transport, in *Calcium and Its Role in Biology,* H. Sigel, ed., Metal Ions in Biological Systems, vol. 17, Marcel Dekker, New York, pp. 187-214.
Fiskum, G., 1985, Intracellular levels and distribution of Ca^{2+} in digitonin-permeabilized cells, *Cell Calcium* **6:**25-37.
Fiskum, G., G. Becker, and A. L. Lehninger, 1980, Regulation of free Ca^{2+} by mitochondrial and endoplasmic reticulum Ca^{2+} influx and efflux, *Fed. Proc.* **39:**900.
Fiskum, G., and R. S. Cockrell, 1985, Uncoupler-stimulated release of Ca^{2+} from Ehrlich ascites tumor cell mitochondria, *Arch. Biochem. Biophys.* **240:**723-733.
Fiskum, G., and A. L. Lehninger, 1981, Mitochondria are both regulators and targets of intracellular Ca^{2+}, *J. Cell Biol.* **91:**286a.
Fiskum, G., and A. L. Lehninger, 1982, Mitochondrial regulation of intracellular calcium, in *Calcium and Cell Function* vol. 2, W. Y. Cheung, ed., Academic Press, New York, pp. 39-80.
Fiskum, G., and A. Pease, 1984, Comparison of mitochondrial Ca^{2+} transport in normal liver and AS-30D hepatoma cells, *Fed. Proc.* **43:**2003.
Fiskum, G., D. Pfeiffer, K. M. Broekemeir, and B. Baroody, 1985, Calcium buffering characteristics and phospholipase activities of rat brain mitochondria, *Biophys. J.* **47:**413a.
Fleschner, C. R., A. P. Martin, M. L. Vorbeck, J. R. Darnold and J. W. Long, Jr., 1983, Ca^{2+} release from energetically coupled tumor mitochondria, *Biochem. Biophys. Res. Commun.* **115:**430-436.
Floridi, A., and A. L. Lehninger, 1983, Action of the antitumor and antispermatogenic agent lonidamine on electron transport in Ehrlich ascites tumor mitochondria, *Arch. Biochem. Biophys.* **226:**73-83.
Floridi, A., M. G. Paggi, S. D'Atri, C. DeMartino, M. L. Marcante, B. Silvestrini, and A. Caputo, 1981, Effect of lonidamine on the energy metabolism of Ehrlich ascites tumor cells, *Cancer Res.* **41:**4661-4666.
Gots, R. E., F. A. Gorin, and S. P. Bessman, 1972, Kinetic enhancement of bound hexokinase activity by mitochondrial respiration, *Biochem. Biophys. Res. Commun.* **49:**1249-1255.
Greenhouse, W. V. V., and A. L. Lehninger, 1976, Occurence of the malate-aspartate shuttle in various tumor types, *Cancer Res.* **36:**1392-1396.
Greenhouse, W. V. V., and A. L. Lehninger, 1977, Magnitude of malate-aspartate reduced nicotinamide adenine dinucleotide shuttle activity in intact respiring tumor cells, *Cancer Res.* **37:**4173-4181.
Hansford, R. G., and A. L. Lehninger, 1973, Active oxidative decarboxylation of malate by mitochondria isolated from L-1210 ascites tumor cells, *Biochem. Biophys. Res. Commun.* **51:**480-486.
Hayashi, J., O. Gotoh, and Y. Tagashira, 1980, A general characteristic of tumor mitochondria: Leakage of endogenous Mg^{2+} on incubation with uncoupler and resultant reduction of uncoupler-stimulated ATPase activity, *Arch. Biochem. Biophys.* **205:**27-35.
Howatson, A. F., and A. W. Ham, 1955, Electron microscope study of sections of two rat liver tumors, *Cancer Res.* **15:**62-69.
Kaschnitz, R. M., Y. Hatefi, and H. P. Morris, 1976, Oxidative phosphorylation properties of mitochondria isolated from transplated hepatoma, *Biochim. Biophys. Acta* **449:**224-235.

Knox, W. E., M. L. Horowitz, and G. H. Friedell, 1969, The proportionality of glutaminase content to growth rate and morphology of rat neoplasms, *Cancer Res.* **29:**669–880.

Knox, W. E., G. C. Tremblay, B. B. Spanier, and G. H. Friedell, 1967, Glutaminase activities in normal and neoplastic tissues of the rat, *Cancer Res.* **27:**1456–1458.

Knowles, A. F., 1982, Differential effects of 2,4-dinitrophenol and valinomycin (+ K^+) on uncoupler-stimulated ATPase of human tumor mitochondria, *Biochim. Biophys. Acta* **681:**62-71.

Knowles, A. F., 1984, Characteristics of adenine nucleotide fluxes and transport in human tumor mitochondria, *Biochim. Biophys. Acta,* **764:**203-212.

Knowles, A. F., and N. O. Kaplan, 1980, Oxidative phosphorylation and ATPase activities of human tumor mitochondria, *Biochim. Biophys. Acta* **590:**170–181.

Kovacevic, Z., 1971, The pathway of glutamine and glutamate oxidation in isolated mitochondria from mammalian cells, *Biochem. J.* **125:**757-763.

Kovacevic, Z., 1972, Possibility for the transfer of reducing equivalents from the cytosol to the mitochondrial compartment in Ehrlich ascites tumor cells by the malate-aspartate shuttle, *Eur. J. Biochem.* **25:**372-378.

Kovacevic, Z., 1974, Properties and intracellular localization of Ehrlich ascites tumor cell glutaminase, *Cancer Res.* **34:**3403–3407.

Kovacevic, Z., and H. P. Morris, 1972, The role of glutamine in the oxidative metabolism of malignant cells, *Cancer Res.* **32:**326–333.

Kurokawa, M., S. Oda, E. Tsubotani, H. Fujiwara, K. Yokoyama, and S. Ishibashi, 1982, Characterization of hexokinase isoenzyme types I and II in ascites tumor cells by an interaction with mitochondrial membrane, *Mol. Cell Biochem.* **45:**151-157.

Lampidis, T. J., S. D. Bernal, I. C. Summerhayes, and L. B. Chen, 1983, Selective toxicity of rhodamine 123 in carcinoma cells in vitro, *Cancer Res.* **43:**716–720.

LaNoue, K. F., and A. C. Schoolwerth, 1979, Metabolite transport in mitochondria, *Ann. Rev. Biochem.* **48:**871-922.

LaNoue, K. F., J. G. Hemington, T. Ohnishi, H. P. Morris and J. R. Williamson, 1974, Defects in anion and electron transport in Morris hepatoma mitochondria, in *Hormones and Cancer,* K. W. McKerns, ed., Academic Press, New York, pp. 131-167.

Lau, B. W. C., and S. H. P. Chan, 1984, Efflux of adenine nucleotides in mitochondria from rat tumor cells of varying growth rates, *Cancer Res.* **44:**4458–4464.

Lau, B. W. C., L. Weber, M. Maggio, and S. H. P. Chan, 1984, Elevated content of cholesterol affects adenine nucleotide transport in tumor mitochondria, *Fed. Proc.* **43:**1876.

Lazo, P. A., 1981, Amino acids and glucose utilization by different metabolic pathways in ascites-tumor cells, *Eur. J. Biochem.* **117:**19-25.

Lazo, P. A., and A. Sols, 1980, Pyruvate dehydrogenase complex of ascites tumor: Activation by AMP and other properties of potential significance in metabolic regulation, *Biochem. J.* **190:**705-710.

Linder-Horowitz, M., W. E. Knox, and H. P. Morris, 1969, Glutaminase activities and growth rates of rat hepatomas, *Cancer Res.* **29:**1195–1199.

Macara, I. G., 1985, Oncogenes, ions and phospholipids, *Am. J. Physiol.* **248:**C3–C11.

McIntyre, H. J., and F. L. Bygrave, 1974*a*, The role of mitochondria in modifying the cellular ionic environment: Calcium-induced respiratory activities in mitochondria isolated from various tumor cells, *Biochem. J.* **144:**551–558.

McIntyre, H. J., and F. L. Bygrave, 1974*b*, Retention of calcium by mitochondria isolated from Ehrlich ascites tumor cells, *Arch. Biochem. Biophys.* **165:**744–748.

Mitchell, P., 1966, Chemiosmotic coupling in oxidative and photosynthetic phosphorylation, *Biol. Rev.* **41:**455-502.

Modica-Napolitano, J. S., M. J. Weiss, L. O. Chen, and J. R. Aprille, 1984, Rhodamine 123 inhibits bioenergetic function in isolated rat liver mitochondria, *Biochem. Biophys. Res. Commun.* **118:**717-723.

Moreadith, R. W., and G. Fiskum, 1984, Isolation of mitochondria from ascites tumor cells permeabilized with digitonin, *Anal. Biochem.* **137:**360-367.

Moreadith, R. W., and A. L. Lehninger, 1984*a,* The pathways of glutamate and glutamine oxidation by tumor cell mitochondria: Role of mitochondrial $NAD(P)^+$-dependent malic enzyme, *J. Biol. Chem.* **259:**6215-6221.

Moreadith, R. W., and A. L. Lehninger, 1984*b,* Purification, kinetic behavior and regulation of $NAD(P)^+$ malic enzyme of tumor mitochondria, *J. Biol. Chem.* **259:**6222-6227.

Moreno-Sanchez, R., 1983, Inhibition of oxidative phosphorylation by a Ca^{2+}-induced dimunition of the adenine nucleotide translocator, *Biochim. Biophys. Acta* **724:**278-285.

Morita, T., M. Mori, F. Ikeda, and M. Tatibana, 1982, Transport of carbamyl phosphate synthetase I and ornithine transcarbamylase into mitochondria: Inhibition by rhodamine 123 and accumulation of enzyme precursors in isolated hepatocytes, *J. Biol. Chem.* **257:**10547-10550.

Muhammed, H., T. Ramasarma, and C. K. R. Kurup, 1982, Inhibition of mitochondrial oxidative phosphorylation by adriamycin, *Biochim. Biophys. Acta* **722:**43-50.

Nicchitta, C. V., and J. R. Williamson, 1984, Spermine: A regulator of mitochondrial calcium cycling, *J. Biol. Chem.* **259:**12978-12983.

Paradies, G., F. Capuano, G. Palombini, T. Galeotti, and S. Papa, 1983, Transport of pyruvate in mitochondria from different tumor cells, *Cancer Res.* **43:**5068-5071.

Parlo, R. A., and P. S. Coleman, 1984, Enhanced rate of citrate export from cholesterol-rich hepatoma mitochondria: The truncated Krebs cycle and other metabolic ramifications of mitochondrial membrane cholesterol, *J. Biol. Chem.* **259:**9997-10003.

Parry, D. M., and P. L. Pedersen, 1983, Intracellular localization and properties of particulate hexokinase in the Novikoff ascites tumor: Evidence for an outer mitochondrial membrane location, *J. Biol. Chem.* **258:**10904-10912.

Pedersen, P. L., 1978, Tumor mitochondria and the bioenergetics of cancer cells, *Prog. Exp. Tumor Res.* **22:**190-274.

Pedersen, P. L., and A. P. Morris, 1974, Uncoupler-stimulated adenosine triphosphatase activity. Deficiency in intact mitochondria from Morris hepatomas and ascites tumor cells, *J. Biol. Chem.* **249:**3327-3334.

Pedersen, P. L., J. W. Greenwalt, B. Reynafarje, J. Hullihen, G. L. Decker, J. W. Soper, and E. Bustamante, 1978, Preparation and characterization of mitochondria and submitochondrial particles of rat liver and liver-derived tissues, *Methods Cell Biol.* **20:**411-481.

Pleshkewych, A., T. C. Maurer, and C. W. Porter, 1983, Ultrastructural changes in the mitochondria of intestinal epithelium of rodents treated with methylglyoxal-bis(guanylhydrazone), *Cancer Res.* **43:**646-652.

Pleshkewych, A., D. L. Kramer, E. Kelly, and C. W. Porter, 1980, Independence of drug action on mitochondria and polyamines in L1210 leukemia cells treated with methylglyoxal-bis(guanylhydrazone), *Cancer Res.* **40:**4533-4540.

Pollakis, G., E. Goormaghtigh, and J. M. Ruysschaert, 1983, Role of the quinone structure in the mitochondrial damage induced by antitumor anthracyclines: Comparison of adriamycin and 5-iminoduanorubicin, *FEBS Lett.* **155:**267-272.

Prentki, M., D. Janjic, and C. B. Wollheim, 1983, The regulation of extramitochondrial steady state free Ca^{2+} concentration by rat isulinoma mitochondria, *J. Biol. Chem.* **258:**7597-7602.

Racker, E., 1972, Bioenergetics and the problem of tumor growth, *Am. Sci.* **60:**56-63.

Racker, E., J. H. Johnson, and M. T. Blackwell, 1983, The role of ATPase in glycolysis of Ehrlich ascites tumor cells, *J. Biol. Chem.* **258:**3702-3705.

Reitzer, L. J., B. M. Wice, and D. Kennell, 1979, Evidence that glutamine, not sugar, is the major energy source for cultured HeLa cells, *J. Biol. Chem.* **254:**2669-2676.

Reynafarje, B., and A. L. Lehninger, 1973, Ca^{2+} transport by mitochondria from L1210 mouse ascites tumor cells, *Natl. Acad. Sci. (USA) Proc.* **70:**1744-1748.

Sauer, L. A., and R. T. Dauchy, 1978, Identification and properties of the nicotinamide dinucleotide $(phosphate)^{+}$-dependent malic-enzyme in mouse ascites tumor mitochondria, *Cancer Res.* **38:**1751-1756.

Sauer, L. A., R. T. Dauchy, W. O. Nagel, and H. P. Morris, 1980, Mitochondrial $NAD(P)^{+}$-dependent malic enzyme activity and malate-dependent pyruvate formation are progression-linked in Morris hepatomas, *J. Biol. Chem.* **255:**3844-3848.

Scarpa, A., and P. Graziotti, 1973, Mechanisms for intracellular calcium regulation in heart. I. Stopped-flow measurements of Ca^{2+} uptake by cardiac mitochondria, *J. Gen. Physiol.* **62:**756-772.

Scholnick, P., D. Lang, and E. Racker, 1973, Regulatory mechanisms in carbohydrate metabolism. IX. Stimulation of aerobic glycolysis by energy-linked ion transport and inhibition by dextran sulfate, *J. Biol. Chem.* **248:**5175-5182.

Schoolwerth, A.C., B. L. Nazar, and K. F. LaNoue, 1978, Glutamate dehydrogenase activation and ammonia formation by rat kidney mitochondria, *J. Biol. Chem.* **253:**6177-6183.

Schreiber, J. R., W. X. Balcavage, H. P. Morris, and P. L. Pedersen, 1970, Enzymatic and spectral analysis of cytochrome oxidase in adult and fetal rat liver and Morris hepatoma 3924A, *Cancer Res.* **30:**2497-2501.

Selkirk, J. K., J. C. Elwood, and H. P. Morris, 1971, Study on the proposed role of phospholipid in tumor cell membrane, *Cancer Res.* **31:**27-31.

Siess, E. A., S. Nimmannit, and O. H. Wieland, 1976, Kinetic and regulatory properties of pyruvate dehydrogenase from Ehrlich ascites tumor cells, *Cancer Res.* **36:**55-59.

Stanisz, J., B. M. Wice, and D. E. Kennel, 1983, Comparative energy metabolism in cultured heart muscle and HELA cells, *J. Cell Physiol.* **115:**320-330.

Summerhayes, I. C., T. J. Lampidis, S. D. Bernal, J. J. Nadakavukaren, K. K. Nadakavukaren, E. L. Shepherd, and L. B. Chen, 1982, Unusual retention of rhodamine 123 by mitochondria in muscle and carcinoma cells, *Natl. Acad. Sci. (USA) Proc.* **79:**5292-5296.

Suolina, E. M., R. N. Buchsbaum, and E. Racker, 1975, The effect of flavenoids on aerobic glycolysis and growth of tumor cells, *Cancer Res.* **35:**1865-1872.

Thorne, R. F. W., and F. L. Bygrave, 1973, Energy-linked functions of tightly coupled mitochondria isolated from Ehrlich ascites tumor cells, *Cancer Res.* **33:**2562-2567.

Thorne, R. F. W., and F. L. Bygrave, 1974, Inhibition by calcium of adenine nucleotide translocation in mitochondria isolated from Ehrlich ascites tumor cells, *FEBS Lett.* **41:**118-121.

Villalobo, A., and A. L. Lehninger, 1979, The proton stoichiometry of electron transport in Ehrlich ascites tumor mitochondria, *J. Biol. Chem.* **244:**4352-4358.

Villalobo, A., and A. L. Lehninger, 1980*a,* Stoichiometry of H^{+} ejection coupled to electron transport through site 2 in ascites tumor mitochondria, *Arch. Biochem. Biophys.* **205:**210-216.

Villalobo, A., and A. L. Lehninger, 1980*b,* The phosphorylation potentials generated by respiring Ehrlich ascites tumor mitochondria, *Arch. Biochem. Biophys.* **203:**473-482.

Villalobo, A., and A. L. Lehninger, 1980*c*, Inhibition of oxidative phosphorylation in ascites tumor mitochondria and cells by intramitochondrial Ca^{2+}, *J. Biol. Chem.* **255:**2457-2464.

Warburg, O., 1930, *Metabolism of Tumors,* Arnold Constable, London.

Warburg, O., 1956, On the origin of cancer cells, *Science* **123:**309-314.

Wehrle, J. P., and P. L. Pedersen, 1981, Isolation and characterization of mitochondria from malignant cells, in *Methodological Surveys (B): Biochemistry Volume 11: Cancer-Cell Organelles,* R. Reid, C. M. W. Cook, and D. J. Morre, eds., Wiley, New York.

Weinhouse, S., 1972, Glycolysis, respiration and anomalous gene expression in experimental hepatomas, *Cancer Res.* **32:**2007-2016.

Williamson, J. R., R. H. Cooper, S. K. Joseph, and A. P. Thomas, 1985, Inositol trisphosphate and diacylglycerol as intracellular second messengers in liver, *Am. J. Physiol.* **248:**C203-C216.

Index

About the Editor

GARY FISKUM is currently associate professor of biochemistry and emergency medicine at George Washington University Medical Center. He received the B.A. degree from the University of California, Los Angeles (UCLA) and the Ph.D. in biochemistry from St. Louis University. From 1978 to 1981 he was a postdoctoral fellow at Johns Hopkins University. His current research focuses on physiological aspects of mitochondrial bioenergetics and cellular Ca^{2+} homeostasis. He and his collaborators have helped define the mechanisms responsible for mitochondrial Ca^{2+} transport and the modes by which they are regulated. Dr. Fiskum has also made important comparisons between the respiration-dependent activities of mitochondria from normal and malignant tissues.